Mammals
Their Reproductive Biology and Population Ecology

J.R. Flowerdew
Department of Applied Biology
University of Cambridge

Consultant Advisor: Professor M.A. Sleigh

Edward Arnold

First published in Great Britain 1987 by
Edward Arnold (Publishers) Ltd, 41 Bedford Square, London WC1B 3DQ

Edward Arnold (Australia) Pty Ltd, 80 Waverley Road, Caulfield East,
Victoria 3145, Australia

Edward Arnold, 3 East Read Street, Baltimore, Maryland 21202, U.S.A.

British Library Cataloguing in Publication Data

Flowerdew, J.R.
Mammals, their reproductive biology and population ecology.
1. Mammals——Reproduction 2. Mammals——Ecology
I. Title
599.01′6 QL739.2

ISBN 0-7131-2896-8

Text set in 10/11 pt Mallard Medium
by Colset Private Limited, Singapore
Made and printed in Great Britain
by Richard Clay plc, Bungay, Suffolk

Preface

A glance at the 21 orders of mammals listed in the appendix to this volume immediately gives the impression of a very diverse group of animals ranging in shape from the long-nosed anteaters and armoured armadillos to the winged bats, the agile rodents and the flippered seals and whales. Together with this vast array of anatomical variation goes a fascinating range of individual and population life styles, specialists and generalists, adapted to almost all the environments on the earth.

Linking together the 4000 + species of mammal are the common characteristics of homoiothermy, internal fertilization, milk production, suckling the young, elaborate dentition and the possession of hair. The monotremes (e.g. the duck-billed platypus) with the production of shelled eggs and the marsupials (e.g. the kangaroos) with their short uterine gestation are highly specialized, but in the other orders further similarities can be found in the basic reproductive anatomy and physiology from which a great variety of adaptations to the environment have evolved. Even in closely related species with an apparently similar anatomy and physiology the life style, behaviour and population biology may be very different.

In this book I draw on examples from the fascinating array of anatomical, physiological and ecological adaptations found in mammalian reproduction and population ecology. These two aspects of mammalian biology are intimately linked because population increase starts with sexual reproduction and the associated physiological and behavioural adaptations and also because physiological and demographic variables appear to have evolved together in creating the particular life-style of the individual and hence the population.

Chapter 1 provides an overview of male and female reproductive physiology and anatomy, from the common embryonic origin to the intricate controls of gamete production; this leads in Chapter 2 to the patterns of variation and adaptation seen in ovarian cycles, fertilization and pregnancy. These 'building blocks' for population increase are subject to environmental, including behavioural and social, influences as discussed in Chapter 3 and, together with the associated demographic variables of, for example, litter size, age at first reproduction, length of reproductive life, they have jointly been the subject of evolutionary pressures and adaptation. The extent of our understanding of the interaction and evolution of these reproductive and life-history variables is discussed in Chapter 4, so linking reproductive biology and population ecology.

In Chapter 5 the description of mammalian populations provides the background information necessary for the proper understanding of the variety of demography and dynamics as well as the methods of analysis encountered in the study of mammalian populations. Whether and why these populations are stable or unstable, common or rare, is discussed in Chapter 6. This leads in Chapter 7 to a discussion of how the 'external', influences of the environment such as predation, disease and food supply and behavioural and social factors as well as 'internal' factors such as physiology and genetics may affect population regulation or limitation and how they may influence mammalian population dynamics. Finally Chapter 8 gives examples of man's use of and influence on mammalian population dynamics and reproduction through exploitation and conservation.

Population processes are influenced by births, deaths, immigration and emigration and by linking the ecology and adaptation of all these parameters it is intended to show where more research is needed and emphasize how few of the mammals have been the subject of more than a cursory physiological or ecological study.

In writing this book I have benefited greatly from discussions with students and colleagues, and particularly Dr J. Clevedon Brown, in the Department of Applied Biology. I am also indebted to Professor P.A. Jewell for suggestions for improvements to parts of the text and to Professor M.A. Sleigh, for his painstaking care in advising on presentation and content. However, I remain fully responsible for any errors in the text.

Last, but not least, I wish to thank my parents for encouraging my interest in mammals and my long-suffering family for their support and patience during the gestation of the volume.

Cambridge, 1987. J.R.F.

Contents

1

The Mammalian Reproductive Tract – Form and Function

Successful reproduction is essential for population survival. To understand fully those processes that underlie population increase or decrease one needs to understand the various internal and external factors which affect reproductive function as well as those which affect mortality and movement. Reproductive function is a subject area where anatomy, physiology and behaviour are intimately connected and far larger works (e.g. Slater, 1978) than this chapter have been written on one aspect alone. Therefore, in order to fulfil the aims of this book it will be necessary to concentrate here upon the development, form and function of the reproductive tract and external genitalia and particularly on the endocrine and neuroendocrine control of the production of gametes ready for fertilization.

Before mating the male and female must possess mature reproductive organs and external genitalia, capable of producing viable spermatozoa and ova, introducing the spermatozoa into the female reproductive tract and nourishing the developing blastocyst and implanted embryo. The reproductive tract of each species is adapted to perform the functions associated with the production, and transmission or reception, of gametes, and mammals show remarkable variation in these structures. However, before considering this adaptation and variation it is important to emphasize the common developmental origin of all mammalian reproductive systems which arise from an embryonic, female-orientated, precursor.

The development of the reproductive tract

Both the male and female reproductive tract develop from the embryonic system of the indifferent gonad and the associated Mullerian and Wolffian ducts (Wilson, 1978; Wilson, Griffin and George, 1980). In the embryo duplicated Wolffian and Mullerian ducts lead into the urogenital sinus which has associated with it a genital tubercle and vestibular folds. The ductular systems develop or atrophy to produce the male or female system (Fig. 1.1), according to the genetic sex of the gonad unless certain chromosomal anomalies are present (Ohno, 1977). Thus it is normal in mammals to have an XY male and an XX female although mutations in the sex-determining genes and their subsidiary regulating systems can cause XX males and XY females. If a Y chromosome is absent the Mullerian ducts normally develop into the female reproductive tract and the Wolffian ducts atrophy. The differentiation of the Wolffian duct system and the associated

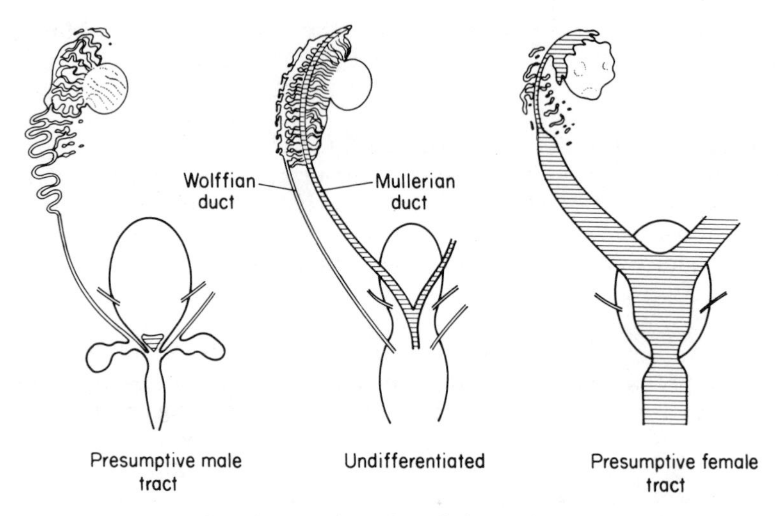

Fig. 1.1 Diagrams to show how male or female genital tracts are derived from the Wolffian or Mullerian duct systems. Redrawn after Hunter (1980).

structures into the male tract depends firstly upon the differentiation of the indifferent gonad into a testis and then the presence of male hormone, testosterone. In the absence of testosterone a female would develop (Jost *et al.*, 1973; Jost, 1979). Femaleness is an inherent trend in the body and would develop in every individual unless foetal testicular hormones impose maleness. The dominance of the male system is also imposed by another testicular hormone (Jost, 1965; Josso, 1974; Josso *et al.*, 1977) called

Table 1.1 Developmental fate of the sexual rudiments in the male and the female mammalian foetus (from Frye, 1967).

Sexual rudiment	Male	Female
Gonad		
Cortex	Regresses	Ovary
Medulla	Testis	Regresses
Mullerian ducts	Vestiges	Uterus, oviducts, parts of vagina
Wolffian ducts	Epididymis, vas deferens	Vestiges
Urogenital sinus	Urethra, prostate Bulbourethral glands	Part of vagina, urethra
Genital tubercle (phallus)	Penis	Clitoris
Vestibular folds	Scrotum	Labia

mullerian-inhibitory factor which induces the regression of the Mullerian duct system. The nature of the changes which occur in the developing male and female reproductive systems are indicated in Table 1.1. Testosterone is also necessary as a substrate for the production of di-hydrotestosterone (DHT) which is responsible for the virilization of the urogenital sinus and the genital tubercle, progenitors of the male urethra and external genitalia (Desjardins, 1981). Testosterone is initially, if not finally, responsible for the masculinization of the brain (Toran-Allerand, 1978) so that masculine patterns of behaviour and physiology are dictated.

The female reproductive system

The female reproductive tract (Fig. 1.2) is physiologically and structurally very malleable, serving to help attract the male during courtship (see Chapter 3), produce ova for fertilization and nourish the embryo. In the non-pregnant state one or both of the ovaries release one or more ova into the peritoneal cavity. The ova are transported by the fimbriated (fringed) funnel of the oviduct towards the uterus. In species such as the rat and the mouse the ovary is enclosed in a sac and is closely applied to the fimbriae; in most other species there is simply an open pouch surrounding the ovary. In cattle this pouch is wide and large, and in pigs it almost encloses the ovary; in horses it is narrow and cleft-like, enclosing only the ovulation fossa of the

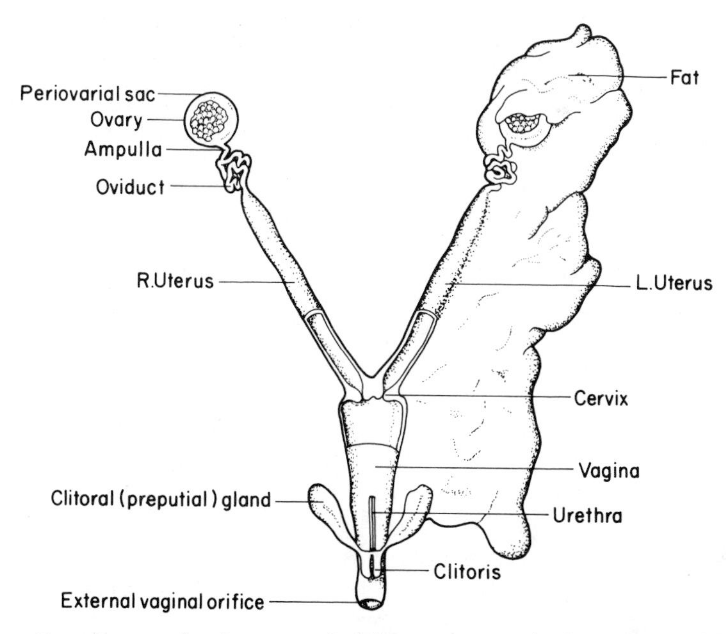

Fig. 1.2 Female reproductive tract of a laboratory rat. The fat is removed from the right side. Parts of the uterine and vaginal walls are cut away to expose the cervices. Redrawn after Turner (1966).

ovary where the ovulating follicles are situated (Stabenfeldt and Hughes, 1977).

The oviduct is divided into three parts; the infundibulum with the fimbriated funnel, the ampulla, where fertilization usually occurs, and the isthmus which leads to the utero-tubal junction. The ampulla runs for half the length of the oviduct and leads to the constricted isthmus. The fimbriae are tentacle-like structures which gather the shed ova or ovum, and may contain ciliated cells which beat towards the uterus so helping the transport of the ovum to the site of fertilization. Further movement of the ovum is usually effected by peristaltic and segmental muscular contractions (Blandau, 1973), until, in many species, ovum transport is prevented by the constriction of the oviduct at the ampullary-isthmic junction. The isthmic part of the oviduct undergoes contractions which help the movement of sperm towards the ovum and later aid the movement of the fertilized ovum to the uterus. The time between ovulation and entry into the uterus may be from 2 to 5 days according to species (Anderson, 1977). In the Llama (*Lama guanicoe*) the isthmus acts as a sperm store (Thibault, 1973). The utero-tubal junction of some species (Fig. 1.3) may control sperm transport in a similar way (Hafez, 1973), providing stores of sperm in pocket-like diverticulae which protrude into the uterine lumen, as in the house mouse, rat, hamster, ferret, dog, mare and pig which show uterine insemination. In the cow, ewe and woman the junction is simple and insemination is intra-

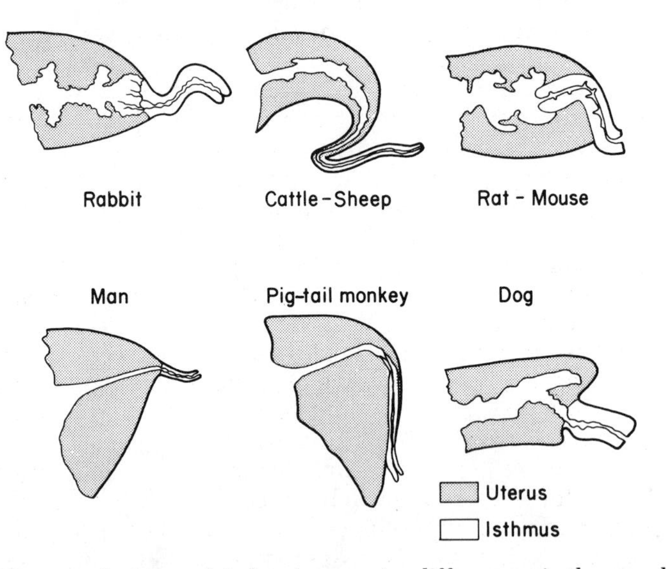

Fig. 1.3 Diagrams (not to scale) showing species differences in the morphology of the uterotubal junction. Note the varied anatomic relationships between the isthmus and the uterus. There are marked folds in the rabbit, a flexure of the isthmus in cattle and sheep, a moundlike papilla in the dog, an archlike junction in the rat and an intramural portion of oviduct in man and the monkey. Redrawn after Hafez (1973).

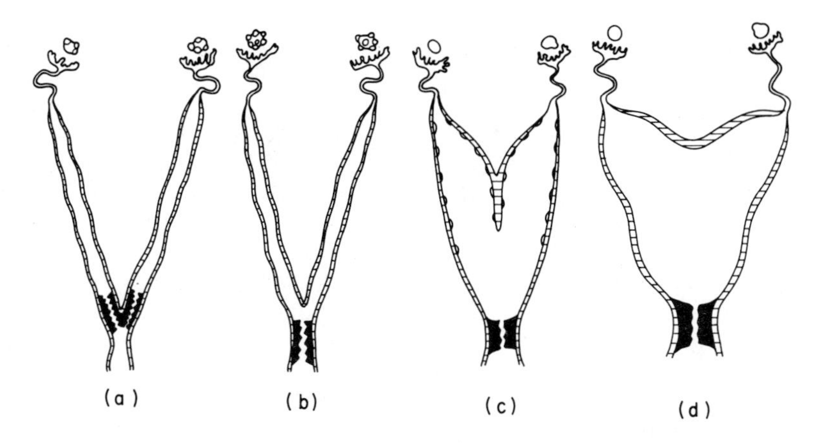

Fig. 1.4 Diagram of the female reproductive tract showing **(a)** the duplex uterus of rodents, **(b)** the bicornuate uterus of the pig, **(c)** the bipartite uterus of ruminants, and **(d)** the simplex uterus of mares and primates. Redrawn from Hunter (1980).

vaginal, however, these relationships do not always follow. In the mouse the utero-tubal junction has been shown to select against sperm with abnormalities (Krzanowska, 1974).

The uterus consists of three layers: an outer visceral peritoneum enclosing a thick layer of smooth muscle, the myometrium, and an inner layer of mucous membrane, the endometrium. This latter layer is vascular and contains tubular uterine glands which act as sperm stores in some species as well as nourishing the embryo. The nature of the layer varies according to the stage of the reproductive cycle (see later). There are two uterine horns, a uterine body and usually one cervix or 'neck' forming the junction between the uterus and the vagina. The shape of the uterus varies with species (Fig. 1.4). Mammals giving birth to many young (polytocous species) are likely to have the primitive duplex uterus, showing the least fusion between the embryonic Mullerian ducts. This has two separate uterine horns and two cervices, as is seen in the rodents and lagomorphs (Eckstein and Zuckerman, 1956). Slightly more fusion occurs giving a small uterine body and single cervix in the bicornuate condition seen in the pig, insectivores, primitive primates, bats and porpoises. The Guinea pig is similar except that the cervix has a single fused caudal opening (os) but separate openings at the cranial end meeting the uterine lumina. Further fusion of the uterus gives the bipartite condition with a moderate-sized uterine body, long horns and a single cervical canal as in the ruminants and carnivores: the uterine body is divided into two halves by a septum. The simplex uterus of most primates shows complete fusion of the uterine horns and body. (Note that the mare is intermediate between the bipartite and simplex condition, retaining the horns but losing the septum.) Obviously litter size is not the only determinant of uterine type but the simplex type is associated with monotocous (single young) species.

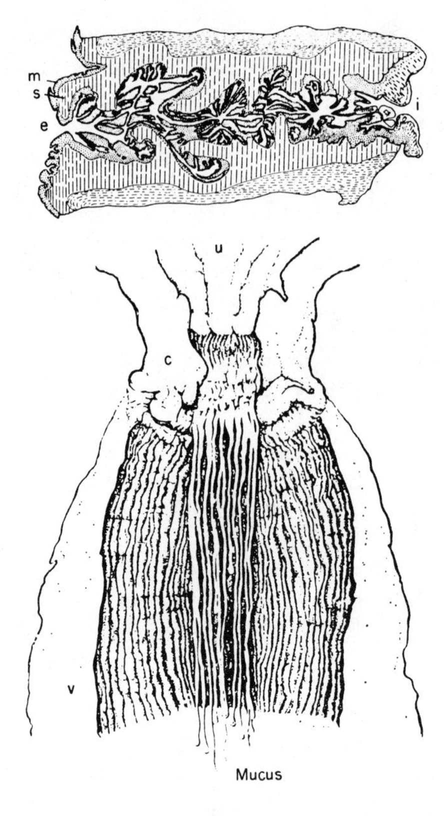

Fig. 1.5 (*Top*) Longitudinal section of the bovine cervix showing the complexity of the cervical crypts: e, external; i, internal; m, mucus-secreting mucosa; s, cervical stroma. (*Bottom*) Diagram of how the strands of cervical mucus flow from the crypts of the cervix, c, at the base of the uterus, u, to the epithelium of the vagina, v. Redrawn from Hafez (1974).

The cervix is usually a sphincter-like fibrous organ (Fig. 1.5), the dilation of which allows passage of the foetus at birth. However, it does possess other functions; in primates, rabbits, cats and many ruminants the cervix produces a mucus secretion which occludes the lumen for most of the oestrous cycle and plugs the cervix during pregnancy (Graham, 1973). The mucus makes the entry of sperm at oestrus (behavioural 'heat' around the time of ovulation) relatively easy as the mucus fibrils change their configuration and allow gaps between themselves so that the sperm can swim along the fibrils. In ruminants and many other species the cervix acts as a sperm reservoir; in the cow and the rabbit the cervix has a complex folded structure within the lumen consisting of mucus secreting crypts. It is into these crypts that the sperm swim when travelling along the mucus fibrils (Mattner, 1966; Morton and Glover, 1974). Subsequently the sperm are slowly released into the uterus and this delay in sperm transport has been suggested to be important for spontaneous ovulators (see Chapter 2) where coitus may not be coincident with ovulation. The barriers to sperm pentration within the female reproductive tract are also considered in Chapter 2.

The vaginal canal and external genitalia of female mammals may not be as obvious as the male organs but nevertheless are very variable between species (Eckstein and Zuckerman, 1956). The clitoris is prominent in carnivores, insectivores, primates and the elephant and may contain a cartilaginous support (cat) or bone (e.g. bear) homologous with the os penis of the male. In the mole (*Talpa europaea*) the clitoris of the breeding female is present as a pendulous prominence and is almost as long as the male's penis (Matthews, 1935). In some species the urethral orifice opens into the urogenital sinus so that a true vagina is absent (e.g. armadillo, *Dasypus* spp.) or the urethra opens into the vagina at its distal end as in the mare, cow, sow and rabbit. In others such as the elephant (*Loxodonta africana*)

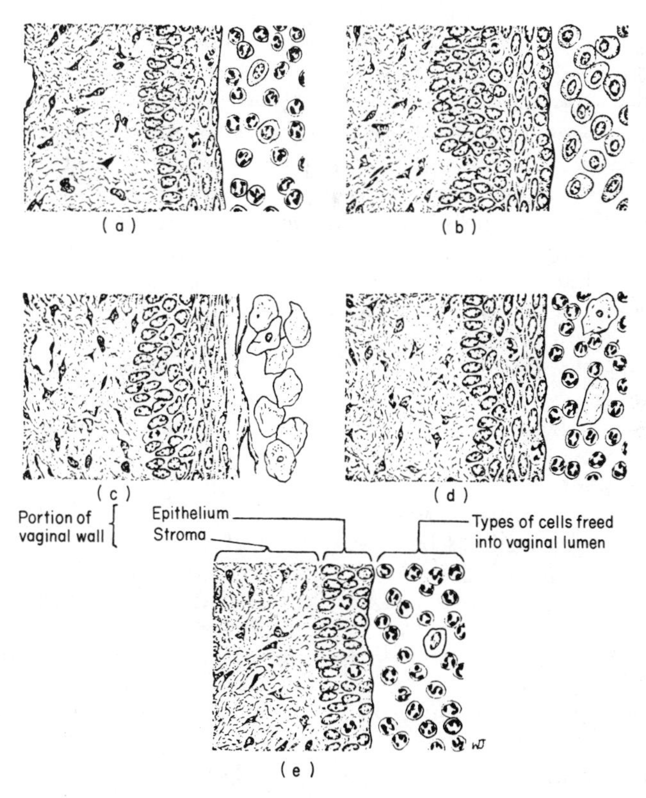

Fig. 1.6 Diagrams of sections through the vaginal wall of the rat during the stages of the oestrous cycle, and showing the types of cells which appear in smears obtained from the vaginal lumen: **(a)** dioestrus; **(b)** pro-oestrus; **(c)** oestrus; **(d)** metoestrus; **(e)** classification of cell layers. Cells with multiple (joined) nuclei are polymorphonuclear leucocytes, other cells are epithelial, either nucleated or cornified, non-nucleated. Redrawn from Turner (1966).

the urethra joins with the vagina to form a long urogenital canal. In many rodents and insectivores the clitoris may incorporate the urethra and so resembles the penis; this arrangement allows separate openings for the urethra, rectum and vagina. The vagina functions as a receptacle for the penis at copulation and it may show marked cyclic fluctuations in its histology during the oestrous cycle; in some rodents the stage of the cycle can be detected by taking a vaginal smear and looking at the proportions of cornified epithelial cells, uncornified epithelial cells and leucocytes coming from the vaginal wall (Fig. 1.6). The external genitalia of many rodents, particularly the vaginal opening, also show cyclical changes in size and coloration (e.g. laboratory mouse (Champlin *et al.*, 1973)). Adult primates such as chimpanzees and mangabeys (*Cercocebus* spp.) possess an area of skin around the perineum which swells up and becomes bright pink during the time of the oestrous cycle around ovulation. The reason why this swelling and coloration occurs in some species but not in others remains obscure (Chalmers, 1979). The vaginal orifice may not be open to the exterior at all times although the female is in breeding condition. In hystricomorphs such as the Guinea pig (*Cavia porcellus*) and the cuis (*Galea musteloides*) there is a vaginal closure membrane which perforates at the time of oestrus so that copulation may take place. Also, in many rodents the vaginal opening closes during pregnancy and reopens prior to parturition and also closes during the anoestrus of the non-breeding season.

The ovary (Fig. 1.7) contains all stages of developing ova from primordial follicles to Graafian follicles which are close to ovulation. Partially developed follicles which have degenerated, called atretic follicles, are also present. Each ovum is surrounded by an envelope of follicle cells which, if ovulation is achieved, becomes invaded by blood vessels and transformed into a corpus luteum which has an endocrinological function in the oestrous cycle and the maintenance of pregnancy. These structures are imbedded in stromal tissue (connective tissue, smooth muscle cells and interstitial cells) and enclosed in a capsule of germinal epithelium (Brambell, 1956).

The ovaries contain a large number of potential ova for fertilization, the number being determined before or soon after birth (Baker, 1982). The primordial germ cells differentiate into primary oocytes and then stop differentiating before the first meiotic division at about the time of birth. No further differentiation occurs until the time of the first ovulation at puberty but further growth does occur. The first meiotic division occurs in most mammals just before ovulation or just after ovulation as in the dog and the fox. The actual number of oocytes present depends on the length of the period of mitotic divisions during oogenesis which occurs in pregnancy and after birth. The oogenetic period continues until degenerative changes start to occur in the germ cells (Mauleon and Mariana, 1977). Oocyte counts then start to fall rapidly as the female mammal grows older; in the cow, out of a population in both ovaries of 120 000 primordial follicles at birth, only about 1000 are left between 15 and 20 years of age (Erickson, 1966).

The development of oocytes to ovarian follicles ready for further maturation and ovulation takes place throughout the life of the mammal (Peters

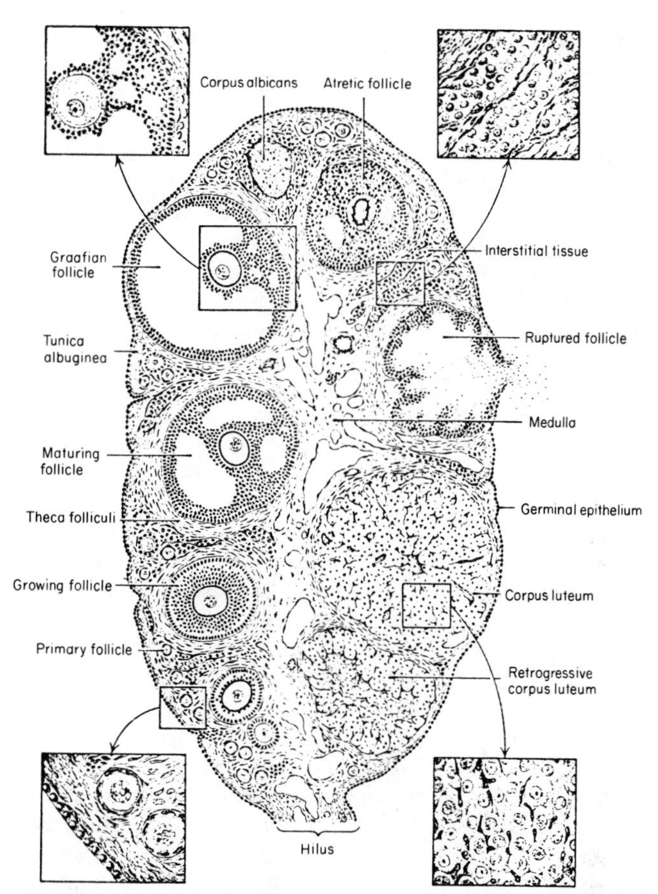

Fig. 1.7 A composite mammalian ovary. The diagrams show the progressive stages in the development of a Graafian follicle (left) an atretic follicle (top) and luteinization following ovulation (right). From Turner (1966).

et al., 1975). The development of the oocytes occurs from infancy onwards and even during pregnancy and seasonal non-reproductive periods. However, only if enough gonadotrophic hormone from the anterior pituitary gland (see p. 11) is present will the oocyte be ovulated and fertilization be possible (note that *intra*-follicular fertilization has been reported for some species such as the shrew, *Blarina brevicauda* (Pearson, 1944)). The actual number of ova ovulated varies between species; in the elephant-shrews such as *Elephantulus myurus* up to sixty ova are ovulated from each ovary but only two young are born (Horst and Gillman, 1940). In the plains viscacha (*Lagostomus maximus*) between 200 and 800 eggs are shed (Weir, 1971).

As the oocyte develops within the follicle (Fig. 1.7) the surrounding epithelial cells of the follicle change from flat to cuboidal and become granulosa cells. Next a 'zona pellucida' is formed between the ovum and the granulosa cells; this is composed of fibrillar material secreted by the granulosa cells and provides a strong protective membrane around the ovum at ovulation. Processes from the granulosa cells traverse the zona pellucida and transfer maternal protein and nutrients to the ovum. Further division of the granulosa cells then occurs and blood vessels invade the outer fibrous layer surrounding the follicle to form the vascular theca interna, leaving the fibrous outer layer of cells as the theca externa. These layers are both responsible for the nutrition of the follicle. At the end of maturation of the ovum the granulosa cells part to form a space, the antrum, which is filled with fluid; the follicle now becomes a Graafian follicle. Only a few cells surround the ovum, being called the cumulus oophorus or corona radiata, protecting the ovum after ovulation. After rupture of the follicle and expulsion of the ovum at ovulation the granulosa cells of the follicle are invaded by blood vessels and further multiplication of these cells produces the corpus luteum.

Endocrine control of the ovarian cycle

Oestrous cycles and the primate modification, menstrual cycles (see Chapter 2), show much variation in detail in their hormonal control but in general terms the mechanisms are often similar. The main difference between species is in the length of life of the corpus luteum which is associated with whether or not it becomes an 'active' endocrine organ. Thus the rat oestrous cycle of 4–5 days and the domestic mammal (sow, cow, mare, ewe) cycle of about 21 days both show follicular growth, ovulation and corpus luteum formation. The latter is short-lived and relatively inactive as an endocrine organ in the rat but long-lived and active as an endocrine organ in the domestic mammals. The reasons for this variation are discussed in Chapter 2.

The oestrous cycle can be divided into two distinct phases; in the first, the follicular phase, culminating in ovulation, the developing follicle secretes the steroid hormone oestradiol-17β, and in the second, the luteal phase, the ruptured follicle is transformed into a corpus luteum secreting another steroid hormone, progesterone. Oestradiol-17β mainly acts on the growth and development of the female reproductive system, promoting the growth of the uterine endometrium and the vaginal epithelium. It also affects secondary sexual characteristics such as the deposition of fat and pelvic enlargement (Jensen, 1979). Progesterone affects the rate of transport of the ovum through the oviduct, prepares the uterus to receive the implanting blastocyst by stimulating the proliferation of the uterine endometrium and inhibits ovulation by blocking the release of pituitary gonadotrophins (see below); it also helps to maintain pregnancy (see p. 14) and stimulates mammary tissue growth.

The decline in progesterone may help to initiate parturition at the end of pregnancy and also, in the menstrual cycle, withdrawal of progesterone initiates the sloughing of the endometrium at menstruation (Heap and Flint, 1979). If fertilization occurs at ovulation then the luteal phase is prolonged and pregnancy ensues; if no fertilization occurs then in polyoestrous species another ovarian cycle follows.

Central to the control of the ovarian cycle are the gonadotrophic hormones follicle stimulating hormone (FSH) and luteinizing hormone (LH) produced by the anterior pituitary gland. FSH stimulates the growth and maturation of ovarian follicles and LH initiates the rupture of the follicle at ovulation and the transformation of the empty follicle into a corpus luteum. The secretion of both of these gonadotrophic hormones and the other anterior pituitary hormones is under the control of regulating hormones or releasing factors produced in the area of the brain immediately above the pituitary gland, called the hypothalamus (Schally, *et al.*, 1973). In this area of the brain neurosecretion of the regulating hormone takes place from axons directly into the portal capillary blood network leading to the anterior pituitary gland (Fig. 1.8), or into the cerebrospinal fluid and via specialized cells (tanacytes) in the median eminence and then into the portal vessels (Karsch, 1984). It was originally thought that luteinizing hormone

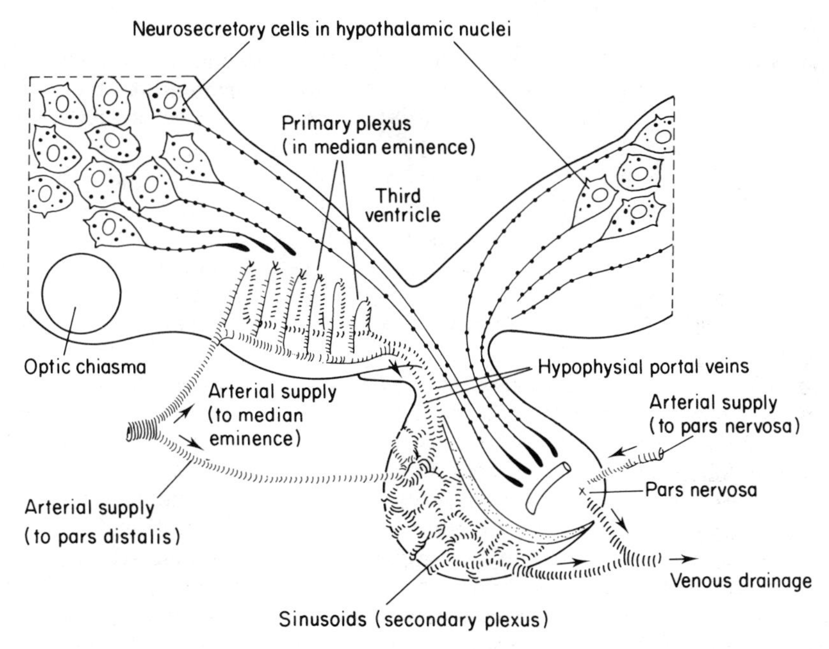

Fig. 1.8 Diagram of the anatomic connections between the hypothalamus and the pituitary gland. Neurosecretory cells in the hypothalamus pass axons down the infundibular stalk to terminate near blood vessels in the pars nervosa; other neurosecretory axons terminate in close proximity to the capillary loops of the median eminence. The hypophysial portal veins pass blood from the primary plexus of the median eminence to the sinusoids of the anterior lobe of the pituitary. The hypothalamic axons of the median eminence liberate gonadotrophic releasing factors directly into the portal vessels and/or into the cerebrospinal fluid in the third ventricle from which they are collected by specialized cells and released close to the portal vessels. After Turner (1966).

releasing factor (LHRF) and follicle stimulating hormone releasing factor (FSHRF) were independently secreted into the portal system but now it is generally accepted that LHRF promotes the secretion of both LH and FSH (Wise *et al.*, 1979) and that the releasing hormone should be called gonadotrophin releasing hormone (GnRH). Further substances of neural origin acting as neurotransmitters, the biogenic amines, also have a role to play in the control of GnRH production (Yen, 1977; Wilson, 1979). The release of GnRH and the gonadotrophins (LH and FSH) occurs in rhythmic pulses (Clarke and Cummins, 1982; see also Lincoln, 1979: Fig. 1.20) which become more frequent at certain times of the ovarian cycle; pulsatile release probably prevents desensitization of the target tissues (Knobil, 1980).

The interaction between the hypothalamic releasing factors, the pituitary gonadotrophins and the gonadal steroids is complex and varies between species. However, the changes in relative levels of the hormones in the blood, for example in the sow, (Fig. 1.9) are similar in other domestic mammals (Hansel and Echternkamp, 1972) and despite differences in detail, the general relationships between the various hormones and their stimulation or inhibition are much the same in those species which have been studied (domestic mammals, laboratory rodents and primates including women). The following account of the endocrine control of the mammalian ovarian cycle is taken mainly from Emmens and Gidley-Baird (1977), Henderson (1979) and Fink (1979).

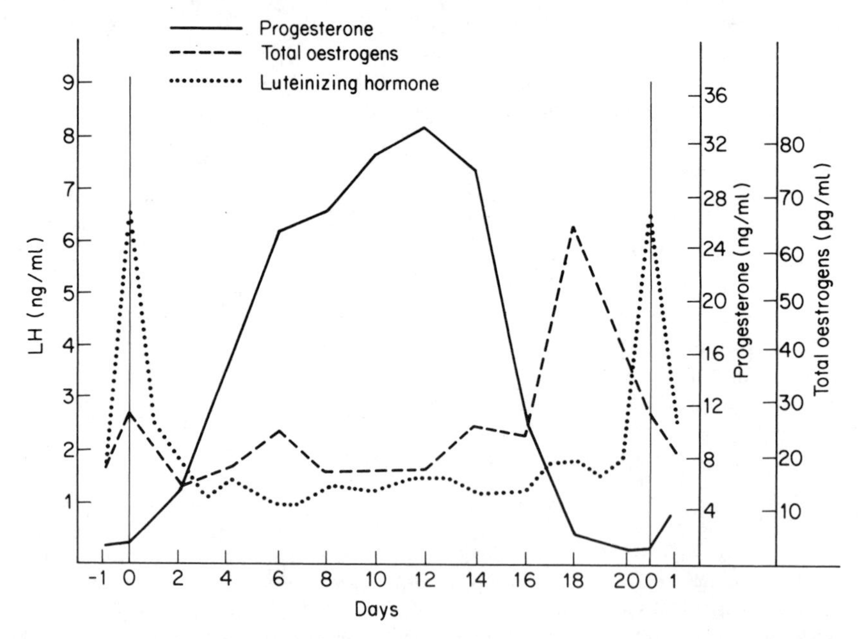

Fig. 1.9 Example of hormonal changes in a domestic mammal, the pig. Peripheral plasma levels of LH, progesterone and oestrogens during the oestrous cycle. Note that 0 on the *x*-axis represents the day of oestrus. Redrawn from Hansel and Echternkamp (1973).

Once oestrogen secretion by ovarian follicles increases at puberty then GnRH production is initiated and this leads to the first LH surge from the pituitary stimulating follicular maturation and ovulation. This stimulation of the follicle occurs by (a) LH binding to the thecal cells of the follicle and influencing the production of extra-follicular androgen (the male hormones testosterone and androstenedione) and oestradiol-17β, and (b) FSH binding to the granulosa cells within the follicle and promoting the aromatization of androgen in the follicle and its conversion to oestradiol-17β. The reduction of the androgen in the follicle is necessary for the successful development of the Graafian follicle and ovulation. Both oestradiol-17β and FSH are retained by the developing follicle and influence its growth when blood FSH levels decline towards the end of the follicular phase of the cycle. The maturing follicle is thus termed 'oestrogenic' in nature in contrast to the other follicles which will not ovulate which are 'androgenic'. These latter follicles do not accumulate enough FSH to reach an ovulatory state and undergo atresia. The FSH and oestradiol-17β in the follicular fluid promotes the formation of more follicular fluid, the proliferation of granulosa cells, the maturation of their ability to synthesize progesterone and the development of LH receptors on the granulosa cells. In some species (e.g. women) prolactin, a hormone produced by the anterior pituitary under the influence of prolactin releasing factor from the hypothalamus, may be necessary for the maturation of the granulosa cells for progesterone production in response to the ovulatory surge of LH. The high levels of oestradiol-17β produced by the follicle induces a positive 'feed-back' effect on the anterior and preoptic areas of the hypothalamus and on the anterior pituitary to stimulate a surge of GnRH, then LH and to some extent FSH at the end of the follicular phase. The LH surge initiates follicular rupture by an uncertain mechanism; proteolytic enzymes and prostaglandins (local hormones) may be involved. The theca externa cells act like muscle cells and the follicle becomes contractile in response to the prostaglandins. At ovulation the ovum is expelled into the peritoneum and high levels of circulating oestrogens cause the period of sexual receptivity or 'heat' in the cycle of many species. This period may occur before, during or after ovulation; thus in the cow oestrus behaviour is usually at an end 10–12 hours before ovulation whereas in the ewe oestrus lasts for about 24–36 hours and ovulation occurs at the end of oestrus 24–26 hours after oestrus behaviour starts.

After ovulation the levels of oestrogen in the blood fall as the follicle changes into a corpus luteum. Gonadotrophin levels fall because progesterone from the corpus luteum feeds back in a negative manner directly to the anterior pituitary and/or via the hypothalamus. Only when the corpus luteum regresses does follicle maturation again continue to ovulation; this is the result of the removal of the progesterone feed-back to the pituitary and hypothalamus and the subsequent increase in GnRH and gonadotrophins.

The luteal phase of the cycle may be relatively long and 'active' in the production of progesterone or short and 'inactive' (see Chapter 2). In the long luteal phase low levels of LH are needed to support the corpus luteum and in the human menstrual cycle the corpus luteum regresses within 7 days without LH. If the ovum or ova shed at ovulation are not fertilized, or a 'pseudopregnancy' (see Chapter 2) is not stimulated, then a new cycle must be initiated and for this the corpus luteum must become inactive and regress. In non-primates the natural 'luteolysin' responsible for the degeneration of the corpus luteum is prostaglandin F2α (PGF2α) produced by the uterus (Baird, 1978). In the primates studied the luteolysin has not been identified but there is evidence to suggest that it may be PGF2α produced by the ovary in response to oestradiol-17β. In both cases the corpus luteum regresses and the progesterone inhibition of GnRH, LH and FSH is removed.

The feed-back mechanisms are complex; thus with oestrogen for example, low levels feed back to the pituitary and to the hypothalamus to decrease the production of GnRH, FSH and LH and maintain threshold levels of these hormones by a negative feed-back effect. However, if follicle maturation is occurring and relatively high levels of oestrogen are present in the blood for a prolonged period of time (e.g. 36 hours for the rhesus monkey, *Macaca mulatta*), then an LH surge follows because of the positive feed-back effect. Thus the feed-back mechanisms depend upon many factors including the timing and duration of exposure of the target organ or organs to the hormone, and also the level of the hormone in the blood circulation. Neural signals are integrated into the cycle and this is well demonstrated in the initiation of the LH surge. Oestradiol-17β stimulates the neural signal for GnRH production and more LH production (positive feed-back), and also induces the increased responsiveness of the pituitary gonadotrophin-producing cells (gonadotrophs) to GnRH. The LH surge reaches a peak when high concentrations of GnRH act on a pituitary which has been made more sensitive to GnRH by the self-priming effect of the hormone and by progesterone secreted during the LH surge. The decline in the LH surge is due to the fall in the portal plasma levels of GnRH and to a lesser extent to a decline in pituitary responsiveness to the hormone; progesterone tends to inhibit the stimulation of GnRH. The neural signal for the production of GnRH may be less important in primates as there is evidence from the rhesus monkey that there is no need for preoptic connections to the basal hypothalamus for passage of the signal for the LH surge. In contrast to this, in the rat, the connections between the median eminence of the hypothalamus and the preoptic area are necessary for the passage of the neural signal for the LH surge.

If fertilization follows ovulation the pregnancy is maintained by 'luteotrophins' of various kinds, including progesterone of luteal and uterine origin, LH, prolactin and, in women, human chorionic gonadotrophin (hCG) produced by the developing blastocyst. The nature of the varying hormonal mechanisms supporting pregnancy is discussed more fully by Heap and Flint (1983).

The male reproductive system

Successful male reproductive activity is dependent upon the production of viable sperm from the seminiferous tubules within the testes and the functioning of the reproductive tract. This includes the production, from the testicular interstitial cells, of the steroid hormone testosterone which is involved in the development and maintenance of the reproductive organs, courtship behaviour and mating. It is also responsible for development of the male tract from the indifferent reproductive system (see p. 2) and also the masculinization of the brain including the suppression of cyclical LH release which occurs in the female (Slater, 1978). The secondary sexual characteristics of the male mammal, including scent glands in many groups, are also dependent upon testosterone. Other organs which become fully functional at puberty are the erectile intromittent organ, the penis, and the accessory reproductive glands. These latter produce the necessary exocrine secretions for the maintenance of spermatozoa as they travel through the male tract and also for their protection within the female tract.

The testes of mammals are not always held outside the abdomen in a scrotal pouch as in man, the bull and the ram (Fig. 1.10). Abdominal testes are not necessarily maintained at the same temperature as the body owing to counter-current blood flow systems but the generally lower temperature of scrotal testes does not appear to be the reason for migration of these organs (Carrick and Setchell, 1977). The testes are compound tubular glands; the bulk of the organ being composed of complex convolutions of seminiferous tubules which form loops opening at both ends into the rete testis; here many tubules anastomose before joining to form the efferent ducts, the vasa efferentia. The efferent ducts lead into the long coiled and ciliated epididymis, which is closely applied to the testis wall, and connect with the vas deferens that leads to the urethra. Spermatozoa produced in the seminiferous tubules mature within the epididymis and form whole semen when joined by secretions from the accessory glands at ejaculation. The head and middle of the epididymis are important for absorption and secretion of fluid and the tail acts as a storage organ (White *et al.*, 1977).

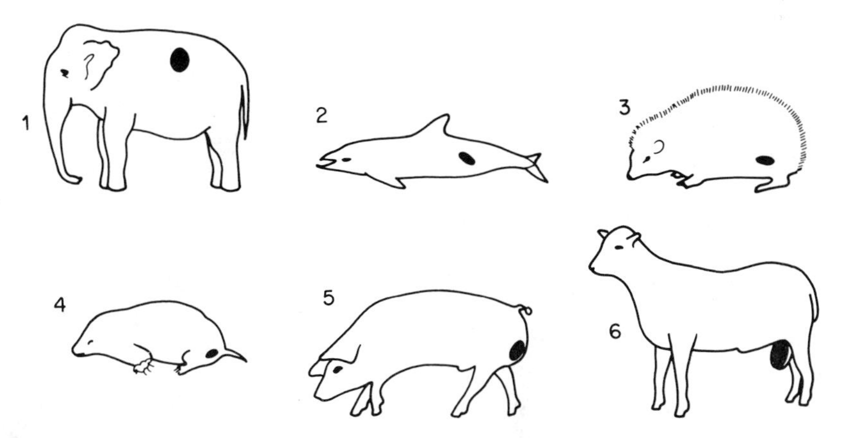

Fig. 1.10 A classification system for the degree of testicular migration in mammals. Types 3 and 6 occur in marsupials and Type 1 occurs in the three extant monotreme genera. Redrawn after Carrick and Setchell (1977).

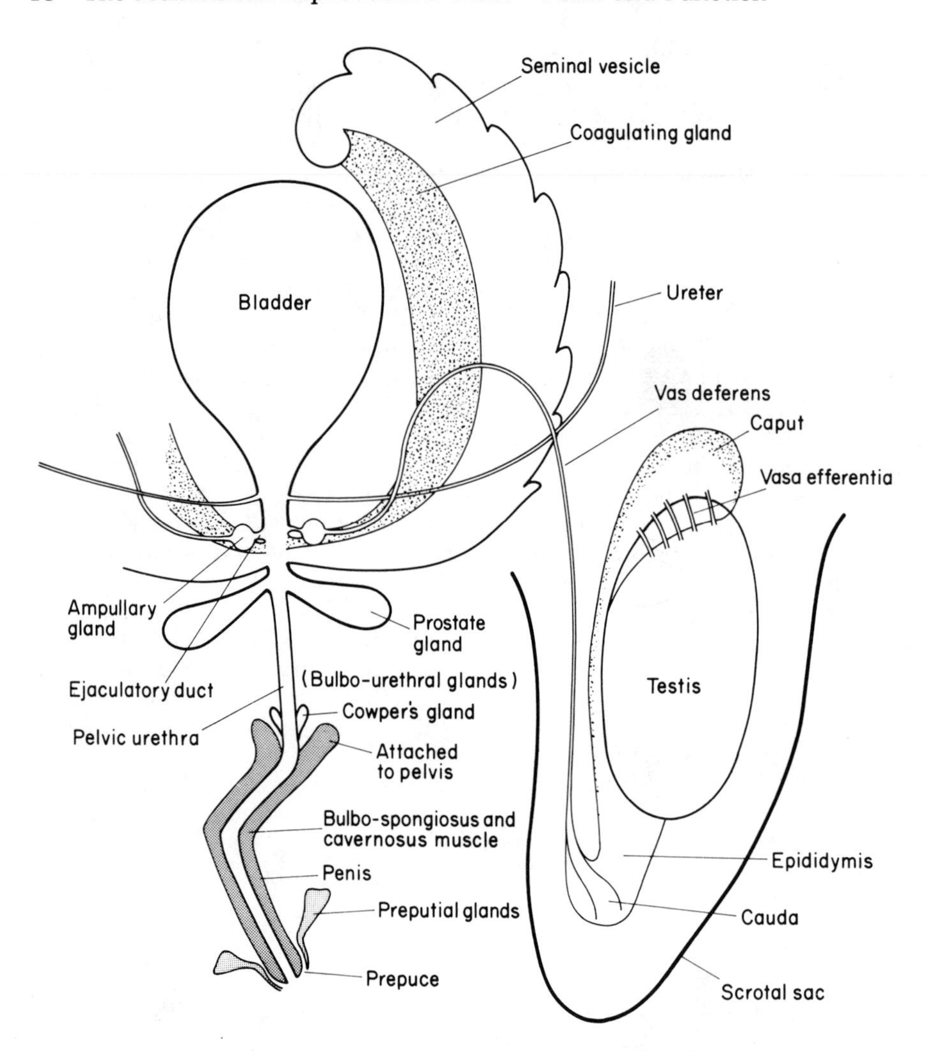

Fig. 1.11 Diagram of the reproductive tract of the male rat.

The anatomy of the male tract varies between species (see Eckstein and Zuckerman, 1956) but the rat (Fig. 1.11) can be taken as typical of many mammals. At the ejaculatory duct the vas deferens joins the urethra and the seminal vesicles make a separate entry (in man the vas deferens and seminal vesicles have confluent openings into the urethra). The nature of the accessory glands is very variable (Fig. 1.12); they may include the ampullary glands, seminal vesicles, prostate glands, bulbo-urethral glands (Cowper's glands) and the urethral glands (glands of Littré). All mammals possess a prostate of some sort; the bull and the ram have a disseminate gland but most mammals have a lobed gland separate from the urethra. The

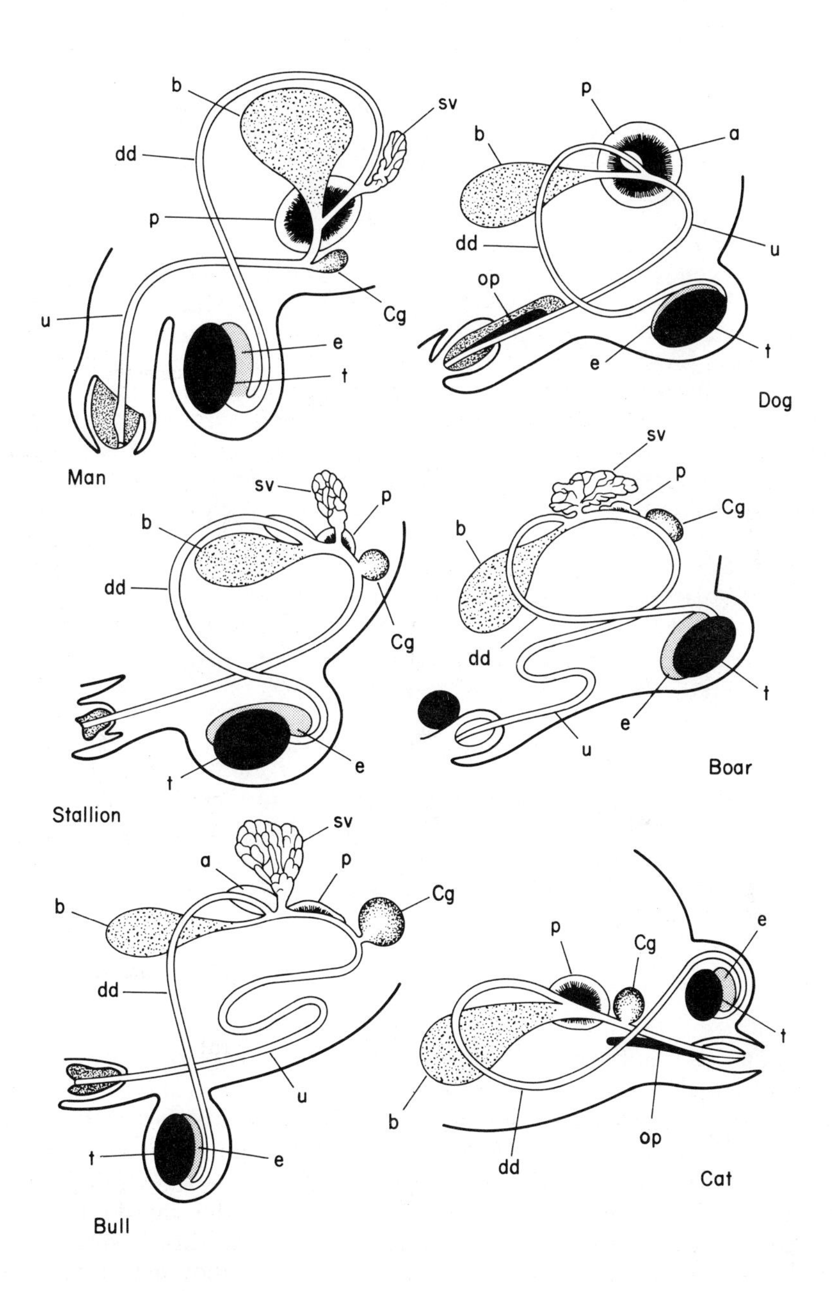

Fig. 1.12 Diagrammatic representation of the reproductive tracts of man, dog, stallion, boar, bull and tom cat: a, ampulla; b, bladder Cg, Cowper's gland; dd, ductus deferens; e, epididymis; op, os penis (baculum); p, prostate gland; sv, seminal vesicle; t, testis; u, urethra. Redrawn after Nalbandov (1964).

functions of the prostate gland and other accessory glands are only partially understood; secretions are very variable between species (Mann, 1964 and 1974; White *et al.*, 1977). The prostate secretes citric acid, acid phosphatase, spermine and zinc in man as well as fructose in the rat. The seminal vesicles are not storage organs for sperm, as their name implies, except in the ram and some deer. In man the secretions of the seminal vesicles include fructose and prostaglandins (biologically active substances acting as local hormones). In the stallion, bull and boar the seminal vesicles produce fructose, citric acid, sorbitol and inositol; in the bull half the ejaculate is derived from these glands. Seminal vesicles are not found in carnivores and cetaceans. The coagulating glands are specialized parts of the prostate gland producing coagulating enzymes; mixing of the fluid from the rat seminal vesicle and coagulating gland (note that they share a common peritoneum but have separate ducts) results in the coagulation of the mixture due to the action of 'vesiculase', an enzyme from the coagulating gland. This process results in the formation of a vaginal plug (see Chapter 4). In the boar the seminal gel is the result of the interaction between the sialomucoprotein of the bulbo-urethral glands and at least two other specific proteins present in the vesicular secretion; another function of the bulbo-urethral glands is the secretion of mucus to neutralize urinary acidity. Urethral glands open into the floor of the urethra and secrete mucus which helps to lubricate the reproductive organs. The glycolytic enzymes of sperm act on the fructose found in the seminal plasma and form lactic acid, providing the energy necessary for the anaerobic survival of the sperm. The various constituents of the seminal plasma secreted from the accessory glands are part of a buffered medium containing easily available substrates which can be metabolized to produce energy (e.g. fructose, glucose and sorbitol), together with bacteriostatic agents and many enzymes. The prostaglandins of human semen have smooth-muscle stimulating properties which possibly accelerate the transport of the semen into the uterus from the vagina.

Male genitalia

Before fertilization is possible the sperm must be introduced into the female reproductive tract via a suitable intromittent organ and then undergo 'capacitation', a process of maturation during which morphological and physiological changes occur so that penetration of the ovum is possible. (See Bedford (1983), Chang and Hunter (1975) for more thorough descriptions of this latter process.) The mammalian penis is very variable in structure and the functions of the various parts are not always fully understood. The comparative anatomy and structure of the penis are given in Eckstein and Zuckerman (1956) and Walton (1960).

The penis, a development of the genital tubercle (see Table 1.1), is attached to the pelvis by ligaments and the muscles at its proximal end. The distal end of the penis is usually concealed within an invagination of the skin of the lower abdomen called the preputial sac or prepuce and the penis only emerges from this position on erection. In species such as the bull the sheath

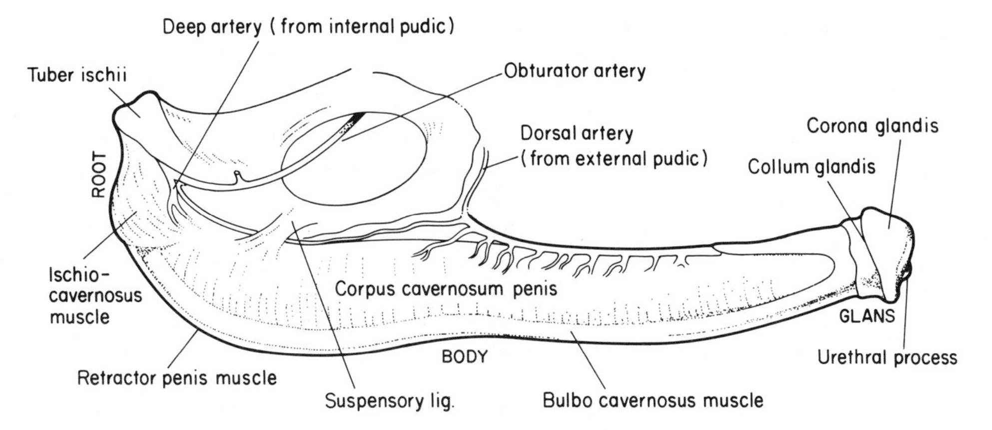

Fig. 1.13 Diagram of the penis of the horse. From Sisson (1975).

is very long and firmly attached to the abdominal wall, completely hiding the penis. In the primates the shealth is shorter and the penis is semi-pendulous in most species; however, in man and tree-shrews and in some bats the penis is truly pendulous, having a free tubular investment of skin which forms a short prepuce around the distal glans. The penis is usually directed anteriorly but may be diverted posteriorly in the non-erect state as in the lagomorphs. The penis of the stallion (Fig. 1.13) is a suitable example of the structure of the organ. It is of the vascular type with large paired corpora cavernosa and a single corpus spongiosum surrouding the urethra; these become engorged with blood during erection. The gland is simple in structure with a dorsal elongation and urethral meatus opening at a urethral process which is slightly off the centre of the glans. Insemination of the mare is intra-uterine. In the bull the penis is of the fibro-elastic type. It is long and slender and even in the non-erect state it is firm and almost rigid. The amount of erectile tissue is small and mainly at the root. As the walls of the erectile tissues are fibro-elastic, rather than muscular as in the vascullar type, the penis becomes slightly more rigid on erection but does not increase in diameter or length. Protrusion of the penis does not simply depend on increased blood flow, as in the vascular type, but also on the relaxation of the retractor penis muscles.

Copulatory behaviour in mammals is variable and complex (Dewsbury, 1972). The nature of the mating behaviour may be related to the type of insemination, intra-uterine or intra-vaginal, and the stimulus which may be necessary for ovulation (see p. 29 and Chapter 4) in induced ovulators or for the ejaculation of semen. In the stallion the penis is kept still after intromission and the preliminary thrusts; the glans swells and distends the orifice of the cervix so the the urethral meatus of the penis is brought into close opposition to the os (opening of the cervix). Spermatozoa are ejaculated in the seminal plasma directly into the uterus. As might be expected, a similar pattern of copulation is seen in the zebra (*Equus burchelli*). In the dog the penis is of the vascular type and coitus is prolonged by the retention

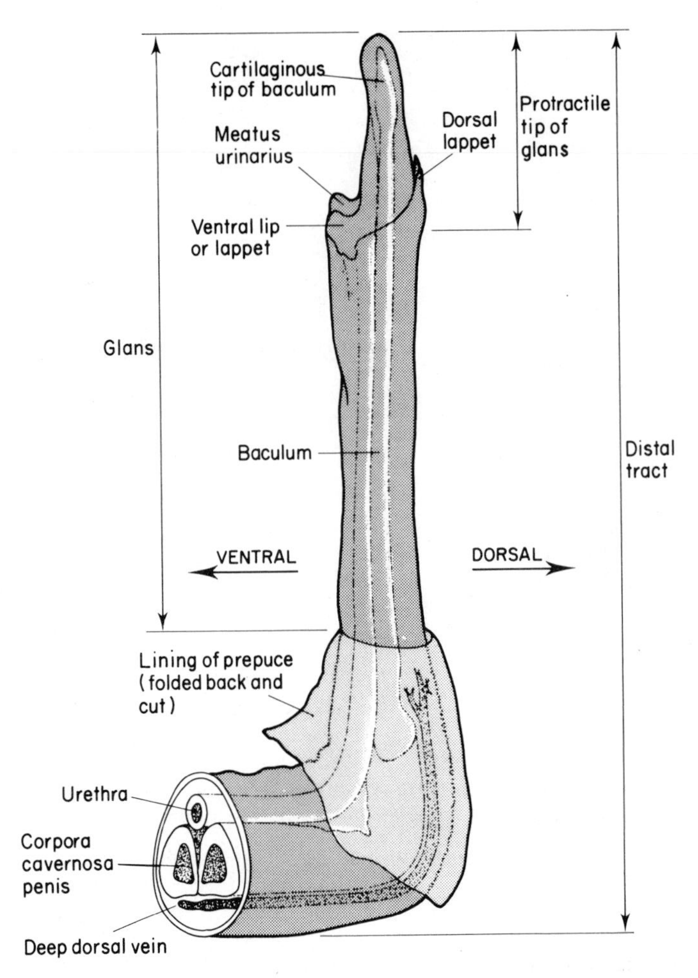

Fig. 1.14 Diagram of the penis of the deer mouse, *Peromyscus maniculatus*. Redrawn after Hooper (1958).

of the penis in the vagina. The base of the glans penis becomes enlarged to form the bulbus glandis which holds the penis in the vagina constricted by muscular spasm. This position, called a 'lock' occurs after pelvic thrusting and dismounting by throwing one leg over the back of the bitch, causing the pair to remain together, back to back. Ejaculation occurs during the lock and insemination is intra-uterine. Such locking patterns of copulatory behaviour are also seen in the fox (*Vulpes vulpes*) and Arctic Fox (*Alopex lagopus*), and many other canids.

In the boar the penis is of the fibro-elastic type and of a small diameter. The boar thrusts until the glans penis, which is cork-screw like, is tightly lodged in the cervix and ejaculation then takes place directly into the uterus.

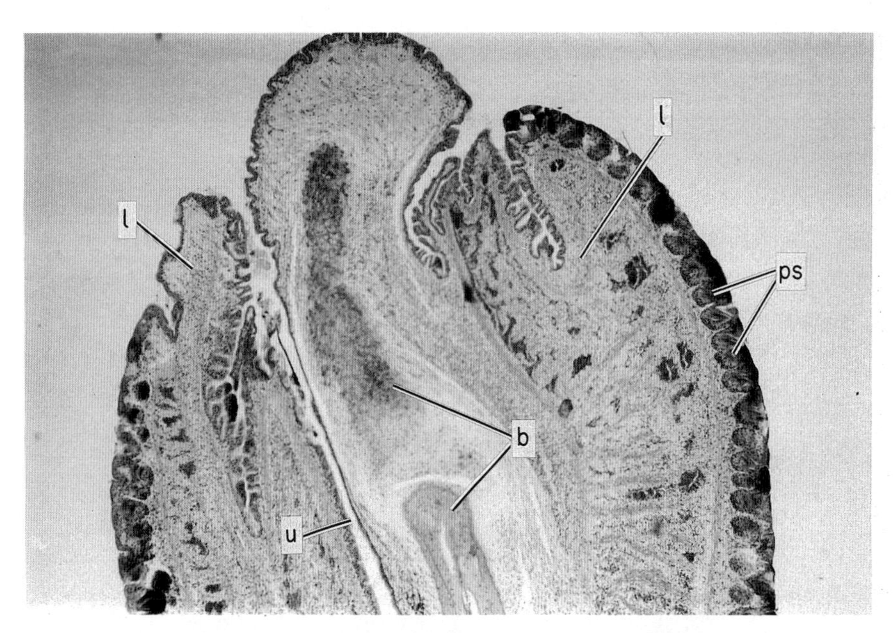

Fig. 1.15 Distal part of glans penis of the bank vole *Clethrionomys glareolus* (× 40). Longitudinal section showing penile spines, ps, in the epithelium. Note also the central bony and cartilaginous baculum, b, and the folded lappets, l, either side of the urethra, u. (*Photo* A. Hilton.)

In the rodents insemination is also intra-uterine but the penis is of an intermediate 'indifferent' type (Fig. 1.14); many rodent species show a lobed and papillated or spiny glans penis (Fig. 1.15). The copulatory pattern usually includes multiple intromissions and multiple ejaculations, as in the rat (*Rattus norvegicus*), mouse (*Mus musculus*), deer mice species of the genus *Peromyscus*, rice rats (*Oryzomys palustris*) and cotton rats (*Sigmodon hispidus*). The function of the penile spines has been debated by Milligan (1979); suggestions include increased male genital sensitivity, provision of additional stimulation for the female, holdfast mechanisms for species with copulatory locks and the facilitation of vaginal plug removal. Neuroendocrine reactions to copulation are necessary for ovulation or more subtle reproductive responses to take place in many rodent species (e.g. McGill and Loughlin, 1970; Diamond, 1970). The stimulus suggestion seems plausible but reflex ovulation can occur in response to simple external stimuli and even a smooth glass rod may be used to induce ovulation in some species. The success of different amounts of vaginal stimulation in the field vole (*Microtus agrestis*) is shown in in Table 1.2.

In the penis of the bull, ram and many artiodactyls the glans penis has a vermiform appendix at the distal end, arising from a point next to the urethral opening. The function of this appendix is uncertain but as ejaculation occurs on withdrawal of the penis and at this time the appendix is trailing, it may help to keep the seminal fluid in close contact with the cervix.

Table 1.2 Induction of ovulation and luteal function in field voles after different amounts of mating. Note that a series of one or more intromissions precedes ejaculation and each series is referred to as an 'ejaculatory series'. From Milligan (1975).

Treatment	Number of females			
	Total	Ovulating	With short-lived corpora lutea	Pregnant or pseudopregnant (functional corpora lutea)
Mounting only				
Parous	6	3	0	3
Virgin	5	0	–	–
Intromission for 5 sec	4	1	0	1
Intromission for 10 sec	4	1	1	0
One complete intromission (mean duration ± S.E. = 22 ± 3.5 sec)	12	11	8	3
One ejaculatory series	6	6	3	3
Two ejaculatory series	6	6	2	4
Four ejaculatory series	10	10	0	10

A further variable throughout the mammals is the presence of a baculum, or os penis. This rigid bony support for the penis is found in most primates, rodents, bats, all carnivores and some insectivores, and it presumably helps to support the erect penis. (Note that females of species where the male has a baculum possess a small bone in the clitoris, the os clitoris).

The prepuce often contains many sebaceous glands and in rodents and some deer it is adapted for scent marking. A positive correlation has been observed between the length of the prepuce in four species of rodent and the frequency of scent marking (Maruniak *el al.*, 1975; Fig. 1.16).

Spermatogenesis and its endocrine control

Sperm production in the seminiferous tubules is a continuous process starting at puberty and only stops if reproduction is seasonal or if physiological or environmental factors inhibit the process in some way (see Chapter 3). The production of sperm is regulated by testosterone produced by the Leydig cells in the interstitial tissue of the testes (Fig. 1.17), which in turn interact with the gonadotrophic hormones. Before the details of this regulatory mechanism are discussed it is necessary to understand the process of spermatogenesis more fully. The following account of spermatogenesis is based on articles by Ewing *et al.* (1980), Dorrington (1979) and Monesi (1972).

Spermatogenesis is dependent upon the division of the germ cells, the spermatogonia, and the physically and physiologically supporting Sertoli cells. Sertoli cells are large in relation to spermatogonia and rest between them on the basement membrane. Adjacent Sertoli cells are joined at specialized 'tight junctions' which prevent movement of blood plasma and

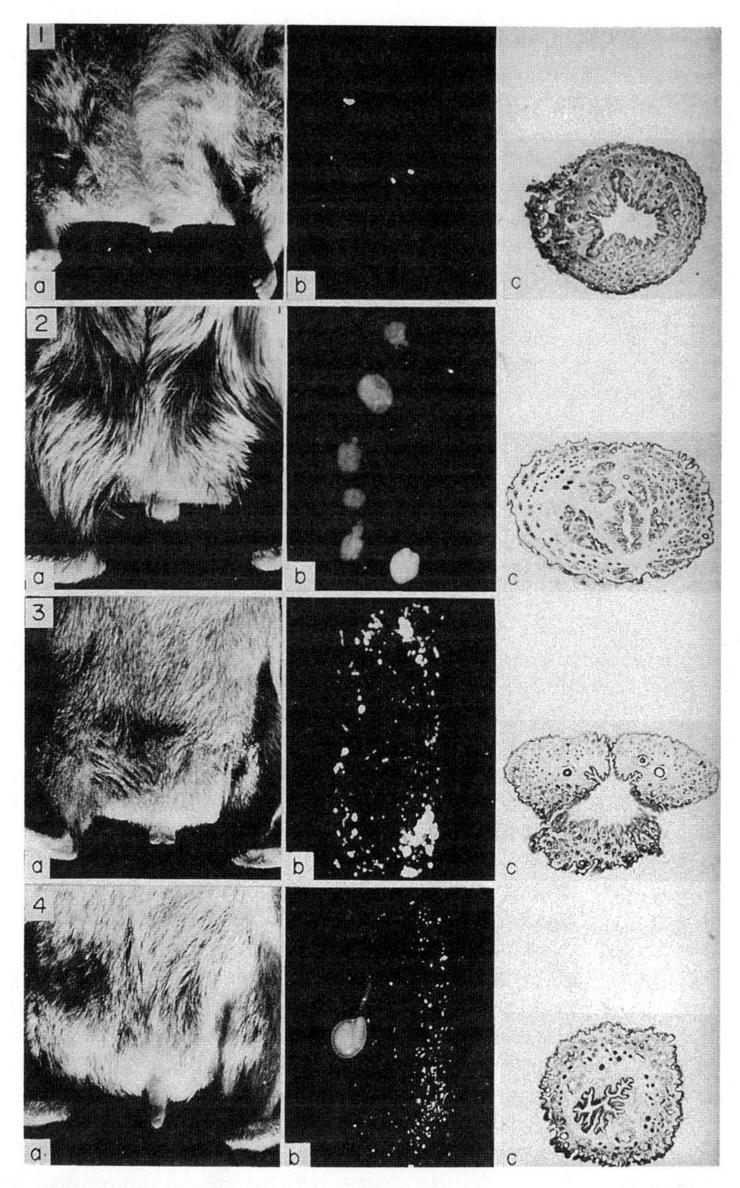

Fig. 1.16 The prepuce and urinary marking patterns in rodents. The photographs show **(a)** the external appearance of the prepuce, **(b)** the typical pattern of urine marking as seen under ultraviolet light, and **(c)** cross-sections of the distal regions of the prepuce of the hamster (1), the Mongolian gerbil (2), the house mouse (3), and the deer mouse (4). Note length of preputial sheath relative to the ankle. (1c = × 9; 2c = × 8; 3c = × 11; 4c = × 12.) From Maruniak, Desjardins and Bronson (1975).

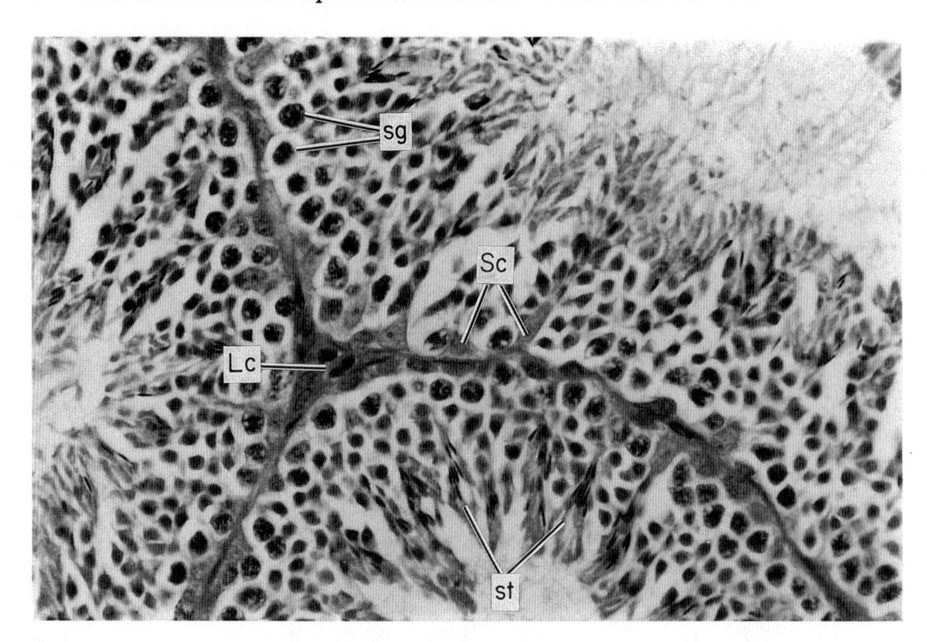

Fig. 1.17 Section (× 140) through mammalian testis (bank vole) showing parts of three seminiferous tubules, Leydig (interstitial) cells, Lc, spermatids, st, spermatogonia, sg, and Sertoli cells, Sc. See text for further explanation. (*Photo* A. Hilton.)

lymph into the lumen of the seminiferous tubule. There is thus a basal compartment confluent with the normal circulatory system and an adluminal compartment providing a milieu suitable for spermatogenesis and particularly meiosis (Fig. 1.18). The developing spermatogonia move adluminally and tight junctions occur above and below so that the compartments are not breached when one junction opens to release the germ cell into the adluminal compartment.

Spermatogenesis starts with the proliferation of spermatogonia by mitotic division so that undifferentiated cells are reproduced together with differentiating spermatogenia. Differentiation proceeds by mitosis to produce spermatocytes which then undergo meiosis and divide to produce the haploid spermatids. Spermiogenesis then follows; this is the transformation of spermatids into spermatozoa which are released into the lumen of the tubule.

Spermatogenesis in rodents occurs in a regular manner so that at a particular time neighbouring segments on the seminiferous tubule contain consecutive stages of spermatogenic development. Thus a series of successive cellular associations (14 associations in the rat) occurs along the length of the tubule and particular stages of development occur in 'waves', having a regular periodicity along the tubule. This is because the differentiation of spermatogenia occurs in a cyclic fashion and the initial mitotic divisions occur at regular intervals (e.g. every 12 days in the rat and 8.6

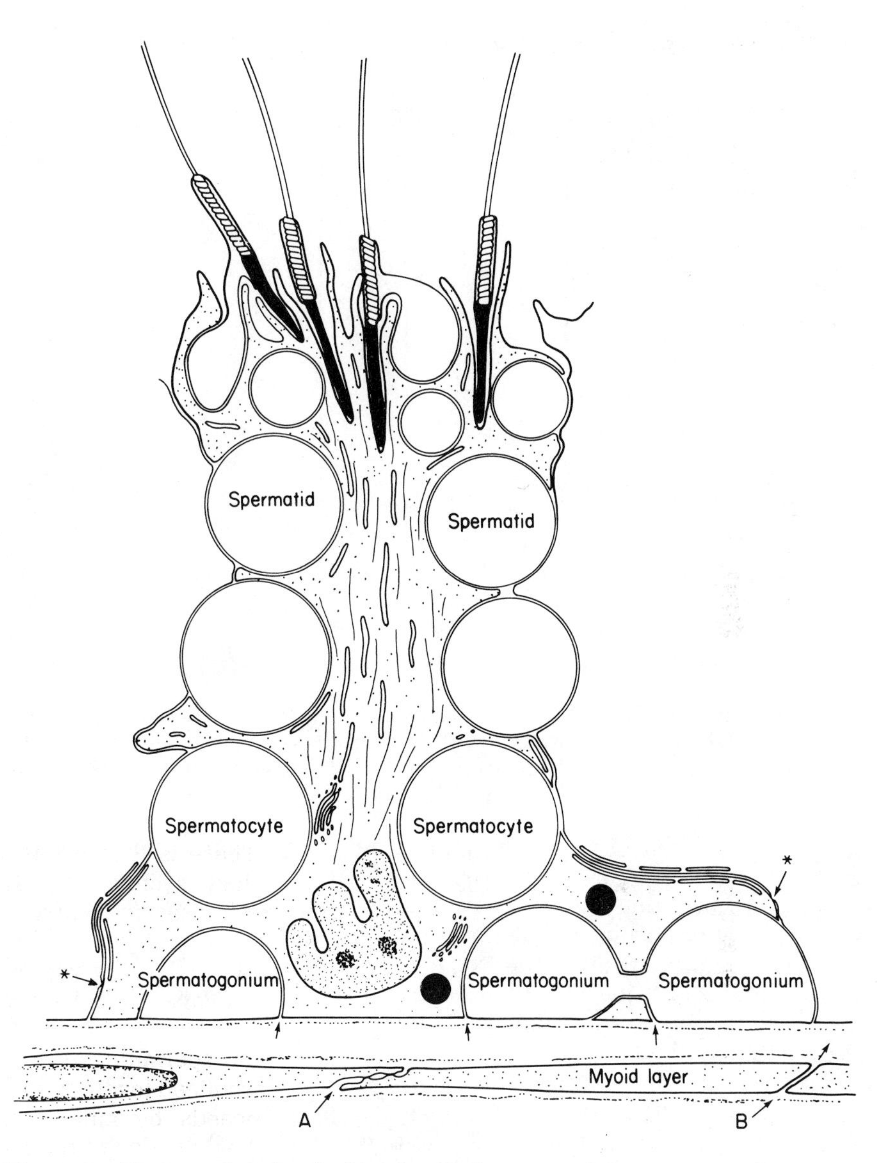

Fig. 1.18 Diagram showing the blood-testis barrier and the compartmentalization of the germinal epithelium by tight junctions between adjacent Sertoli cells. The germ cells are embedded in the columnar Sertoli cell. The primary barrier to substances penetrating from the interstitium is the myoid layer. Most cell junctions in the myoid layer are closed by a tight apposition of membranes as indicated at A, but some are open as depicted at B. Material reaching the base of the epithelium by passing through open junctions in the myoid layer is free to enter the intercellular gap between the spermatogonia and the Sertoli cells in the *basal* compartment. Further penetration is prevented by tight junctions (asterisks) between the Sertoli cells. The tight junctions form a second and more effective component of the blood-testis barrier and to reach the cells in the *adluminal* compartment, substances must pass through the Sertoli cells. Redrawn from Dym and Fawcett (1970).

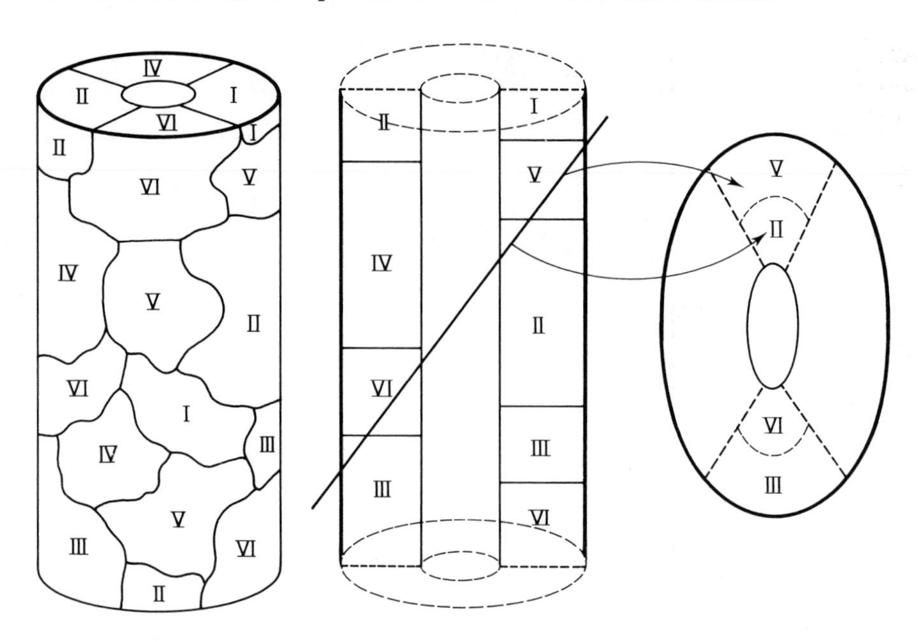

Fig. 1.19 Diagram of a human seminiferous tubule showing the mosaic pattern of distribution of the stages of the spermatogenic cycle (I–VI). This is in contrast to the linear pattern seen in rodents. Redrawn from Heller and Clermont (1964).

days in the house mouse); a section through the testis shows many different stages of spermatogenesis within one tubule as well as in different tubules. This regular ordering of the wave of development along the tubule and the occupancy of particular lengths by particular associations does not occur in every species. In man and some other species an orderly wave of spermatogenesis along the tubule does not occur because different cell associations occupy different portions of the circumference, forming a mosaic pattern (Fig. 1.19).

The process of sperm production is under the control of LH and FSH from the anterior pituitary gland acting together to form an LH-FSH complex (Lincoln, 1979; deKretser *et al.*, 1977). LH acts via the Leydig cells to stimulate testosterone production and FSH acts via the Sertoli cells although the direct effect of this hormone on the germ cells is still a possibility. Pulsatile release of LH in several secretory episodes per 24 hours (Fig. 1.20) stimulates the secretion of testosterone from the Leydig cells, usually within 30 minutes. However, in man there is probably a delay in the response of about 4 hours. The testosterone produced masculinizes target tissues in the body including the testes where androgen receptors are found on the Sertoli cells, the myoid cells of the tubules and possibly on the germ cells. In response to the testosterone stimulation the Sertoli cells produce a number of proteins and steroids including oestrogen, androgen binding protein (ABP) and inhibin and also much of the fluid which is released into the lumen

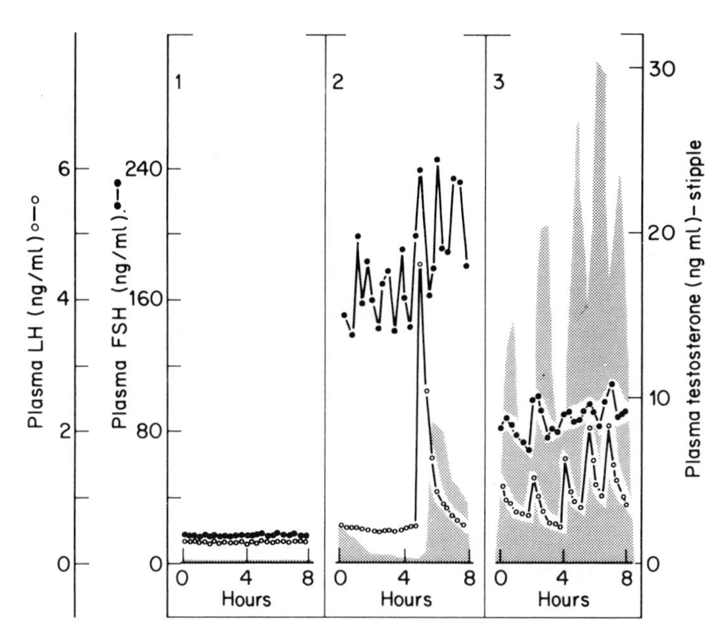

Fig. 1.20 Concentrations of LH, FSH and testosterone during the seasonal sexual cycle of the ram. The diagram shows changes in the concentration of these hormones in the blood plasma of an adult Soay ram sampled at 20-minute intervals for 8 hours during different phases of testicular development: **(1)** testes fully regressed; **(2)** testes redeveloping; and **(3)** testes fully enlarged. Note the increase in frequency of the episodic peaks in LH during reactivation of the testes and the very marked increase in the concentrations of testosterone when the testes are fully developed. Redrawn after Lincoln (1979).

of the tubule. FSH acts synergistically with LH and also independently to promote the division of the germ cells in spermatogenesis. The LH and FSH are stimulated in the same way as in the female mammal, by neural coordination of the secretion of GnRH into the median eminence portal blood. Testosterone feeds back in a negative manner to GnRH production inhibiting LH and FSH release. If testosterone levels are reduced then LH and FSH levels increase. Oestradiol-17β and particularly inhibin (Scott and Burger, 1981) appear to be important in affecting FSH secretion either at the level of the pituitary or via the hypothalamus. The differentiation of spermatogonia takes place in close contact with the Sertoli cell and stimulation of cell division either takes place via the Sertoli cells or possibly directly by the action of testosterone.

It is clear that there are still many facets of both male and female reproduction which are not yet fully understood; the complex nature of the neural and endocrine control of reproduction is illustrated by the fact that male mice release LH when exposed for a short time to a female (Graham and Desjardins, 1980), indicating that a neuroendocrine response can be socially induced.

2

Reproductive Patterns in Mammals

Reproductive physiology and behaviour vary between the many different species of mammals; in this chapter their nature and functions and the extent of variation are described. Some of the information on physiological mechanisms comes from experimental work on the species concerned, but much comes from comparisons with other species. Often comparison, logic, and intuition are all that one has to guide discussions concerning their adaptive functions. These questions have been widely discussed in recent years (Wilson, 1975; Krebs and Davies, 1977) and have led to some intriguing insights into evolution and adaptation. However, it must always be remembered that many generalizations probably come from very little experimental evidence, if any. As the reader will see in this chapter in relation to ovulation, it is quite possible to logically argue that a particular reproductive feature does not show any particular evolutionary trend and later authors argue for the evolution of the feature in two opposite directions! The same is true for adaptation and the reader should bear in mind that the question 'how' is probably far easier to answer than 'why'.

If the main events in reproduction from ovum production to weaning and independence are followed, each step shows significant variation between species and yet the number of species which has been investigated thoroughly at each stage remains small. Weir and Rowlands (1973) point out that exhaustive reproductive studies have been carried out on only 40 or so mammals; they also state that a small suborder, the hystricomorph rodents, appear to be more diversified than any other group of related mammals. It is thus worth remembering that we may be looking at only a few points in a very wide spectrum of variation.

Variation in the ovarian cycle and its control

The production of gametes for fertilization is generally a cyclic process in female mammals (Chapter 1). The ovarian cycle culminates in the shedding of one or more ripe ova at ovulation which is usually fertilized (the non-pregnant cycle is a rarity in the wild (Conaway, 1971)), but if mating is infertile or does not take place at all then polyoestrous (many reproductive cycles) species and possibly dioestrous (two cycles) species will normally initiate another ovarian cycle in order to become pregnant. How quickly a new cycle starts depends upon the nature of the follicular phase, ovulation and the luteal phase of the cycle (p. 10); these three aspects of reproductive physiology all show variation between (and sometimes within) mammalian species.

The types of ovarian cycle

The follicular phase shows some variation in length but its function is always the same: to produce ripe ova for ovulation. Similarly, the luteal phase not only varies in length, it also varies in function, depending upon whether or not the corpus luteum is fully activated as an endocrine organ. If a long and active luteal phase occurs in a normally short (with an inactive luteal phase) non-pregnant cycle then the cycle may become a pseudopregnancy, sometimes lasting as long a normal pregnancy (Conaway, 1971). In some species the luteal phase may become spontaneously active so that it is always comparatively long and the mammal is automatically in a pregnant or pseudopregnant state even if mating has not taken place. Alternatively, the luteal phase may be relatively short in the normal non-pregnant cycle but become induced into a longer, active phase after the stimulus of mating. The ovulation at the end of the follicular phase of the cycle may also be sponteneous or reflexly induced by mating and in the latter case there will not be any luteal phase unless the mating stimulus is strong enough to induce ovulation. There are a few cases where the nature of the luteal phase following induced ovulation varies with the type of mating stimulus. Table 2.1 shows the ways in which these variations are linked together in mammals.

Until recently the classification of ovarian cycles gave only three types (I–III) as described by Conaway (1971). However, Milligan (1974, 1975) discovered type IV in the field vole (*Microtus agrestis*) and this is now incorporated into the classification (Kenny and Dewsbury, 1977) and seems to make the combinations of induced and spontaneous ovulation and luteinization complete. Type I, with a spontaneous ovulation and luteal phase, has a

Table 2.1 Non-pregnant ovarian cycles in mammals.

Type	I	II	III	IV
Ovulation	Spontaneous	Induced	Spontaneous	Induced
Luteal phase	Spontaneous Active	Spontaneous Active	Induced Inactive, short phase unless mated	Induced Inactive, short phase unless sufficient mating stimulus given
Examples	Higher primates	Cat		Field vole (*Microtus agrestis*)
		Ferret	Murids	
	Canids			Montane vole (*Microtus montanus*)
		Microtines		
	Domestic mammals (ungulates)	Lagomorphs Soricids Camelids		

follicular phase of variable length (varying from a few days to several weeks) and may be divided into two sub-groups according to the length of the luteal phase. In ungulates, hystricomorph rodents and the higher primates the cycles are of a medium length, between 2 and 5 weeks, and variation occurs in the follicular phase, as the luteal phase is usually 12–16 days in length. These species are generally polyoestrous. In the second sub-group, including the dog and probably most other canids, the luteal phase lasts from one to two months and, except for the domestic dog and bush dog (*Speothos venaticus*), there is only one period of oestrus, followed by a long period of anoestrous each year (Kleiman, 1968).

In the higher primate ovarian cycle (Type I) the long follicular phase contrasts with that of most other mammals; it is about 14 days long, being approximately the time taken for the gonadotrophin-dependent follicular development. Generally, the follicular phase is much shorter than this, being for example 3 days in the sheep and 7 days in the pig. The explanation of this difference between the higher primates and other mammals lies in the pattern of follicular development. In most mammals the follicular phase of growth from a developing follicle with 4–5 layers of granulosa cells to a mature pre-ovulatory follicle is probably 10–17 days, the same as in the mouse (Pederson, 1970). Baird *et al.* (1975) argue that follicular growth normally progresses smoothly through the luteal phase as long as there is no source of oestradiol other than that from the Graafian follicle. However, if oestradiol from the corpus luteum is present (as in the higher primates) then the GnRH production is suppressed during the luteal phase and follicular growth is impaired. This means that in the higher primates the new follicular phase has to start from an early stage of follicular growth at the time of degeneration of the corpus luteum rather than simply finish off what was started in the luteal phase as in other mammals (Fig. 2.1).

In most of the species showing Type I cycles the young are dependent upon the mother for a relatively long period of time; there is presumably little selective pressure for a shortening of the cycle, either during the follicular phase or during the luteal phase, and this would particularly apply to the higher primates.

The ovarian cycles found in Old World primates, apes and women (see Chalmers, 1979) are called menstrual cycles and differ from other mammals' ovarian cycles by having loss of blood and sloughed-off endometrial tissue from the uterus (menstruation) at the end of the luteal phase and no behavioural 'heat' period at ovulation. Copulation is possible throughout the cycle but tends to occur in many species more frequently around the middle of the cycle when ovulation takes place. However, the difference in cycle type between the primate taxonomic groups is not as clear-cut as these generalizations might suggest; further discussion of the subject may be found in Perry (1971) and Chalmers (1979).

Type II ovarian cycles with induced ovulation and a spontaneous luteal phase may be divided into two subgroups depending on the length of the cycle and the nature of the oestrous period (heat). In the first subgroup the cycles are less than one month in length and oestrus is more or less behaviourally induced (Conaway, 1971). Proximity to a male or continued

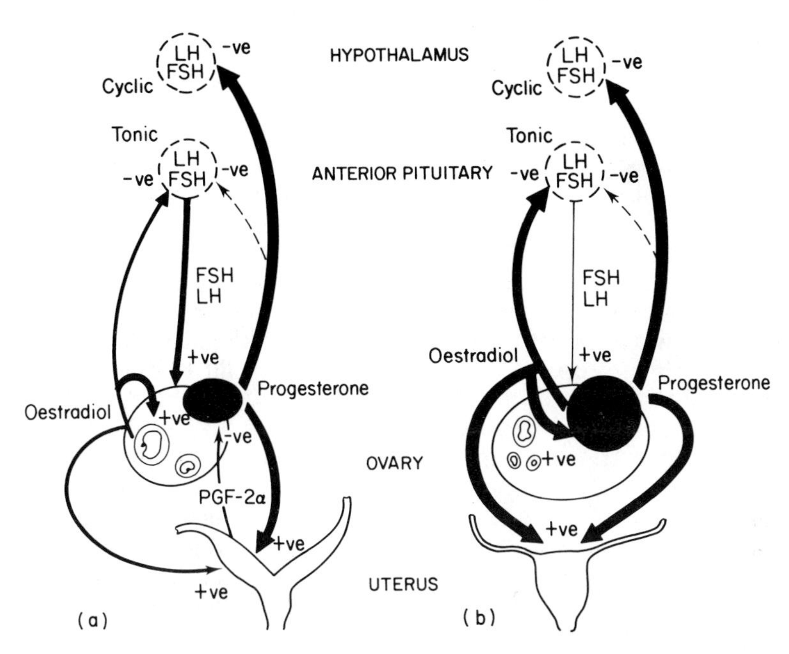

Fig. 2.1 Possible hypothalamic-pituitary-ovarian relationships in the luteal phase of the ovarian cycle in **(a)** sheep and **(b)** women. The magnitude of each effect is related to the thickness of the line. In both sheep and women the positive feedback effect of oestradiol is inhibited by the secretion of progesterone from the corpus luteum (CL). In women the CL controls the basal secretion of gonadotrophins through the negative feedback of oestradiol at the tonic centre. In the ewe the CL does not secrete oestradiol, the basal seretion of gonadotrophins is controlled by the negative feedback of oestradiol from the follicles which continue to develop throughout the luteal phase. The reduction of CL activity at the end of the luteal phase in the ewe depends on an additional negative-feedback loop involving the production of prostaglandin F-2α from the uterus. Redrawn from Baird *et al.* (1975).

social stimulation are often all that is necessary to induce oestrus behaviour. Typical examples are the Microtinae (voles and lemmings) Lagomorpha (rabbits, hares and pikas) and some Insectivora, including the Soricidae (shrews). Unless a mating stimulus occurs oestrus will often continue for many days: up to 30 days of oestrus have been reported for *Microtus ochrogaster*, and ovulation will occur 10.5 hours after mating (Richmond and Conaway, 1969). The second subgroup shows cycles 4–8 weeks in length and oestrus behaviour is more spontaneous and fixed in length than in the previous group. The domestic cat and captive wild felids, with an oestrus of 6–7 days, are typical of this group; they are seasonally or annually polyoestrous (Sadleir, 1966). Also in this group are the ferret (*Mustela furo*), mink (*Mustela vison*), weasel (*Mustela nivalis*) and probably several species of marten (*Martes* spp.) and otter (*Lutra* spp.) (Asdell, 1966; Perry, 1971). A number of unrelated species also probably appear in this subgroup, namely the American mole (*Scalopus*), the American opossum (*Didelphis virginiana*), the camel (*Camelus* spp.), llama (*Lama glama*) and alpaca (*Lama pacos*) (Perry, 1971).

The Type III ovarian cycle shows spontaneous ovulation and a short, inactive, luteal phase unless a longer pseudopregnant luteal phase is induced by infertile mating or a similar stimulus. In the short cycle no functional corpora lutea are formed and ovulation recurs after 4–7 days; the pseudopregnant cycle lasts for 2–5 weeks. This type of cycle is very restricted in its distribution in the orders of mammals; it only occurs in the rodent family Muridae, mainly in the voles, mice and rats.

Type IV ovarian cycles have been described only from the microtine rodents *Micortus agrestis* and *Microtus montanus* (Milligan, 1974, 1975; Kenny and Dewsbury, 1977). In these species mating induces ovulation in the same way as in the Type III cycle but the activation of functional corpora lutea depends on the amount of copulatory stimulation. Following low levels of copulatory stimulation the corpora lutea degenerate after about three days and the active luteal phase is only induced after a more prolonged mating stimulus.

Having divided ovarian cycles into four types it is important to realize that, particularly in the case of induced and spontaneous ovulators, there is no hard and fast dividing line. As Eckstein and Zuckerman (1956) pointed out, the distinction is more 'one of degree rather than kind'; typically spontaneous ovulators such as the rat (*Rattus norvegicus*) and the chinchilla (*Chinchilla laniger*) may, under certain circumstances (such as mating), be induced to ovulate (Weir, 1973; Brown-Grant *et al.*, 1973) and a similar effect has even been reported for man (Eckstein and Zuckerman, 1955). Conversely, induced ovulators may ovulate spontaneously as seen in the rabbit (*Oryctolagus cuniculus*) (Eckstein and Zuckerman, 1956).

Evolution and adaptation of the ovarian cycle

Asdell (1966), described the occurrence of induced and spontaneous ovulation in mammals and decided that 'in view of the apparent sporadic nature of the distribution of the two types of ovulation no conclusions can be drawn regarding evolutionary trends in the character'. However, Everett (1961) and Conaway (1971) believe that because induced ovulation appears to be common in the primitive eutherian order, the Insectivora, and in the primitive (sciuromorph) rodents, and also because induced ovulation is more widespread in mammals than spontaneous ovulation, it seems that induced ovulation is the basic eutherian pattern. On the other hand, Weir and Rowlands (1973) point out that this is unlikely because most other vertebrates, except for some bird species, show spontaneous ovulation.

It is equally easy to argue in two opposite directions in connection with the evolutionary advantage in developing one type of ovulation from another. Conaway (1971) suggests that in ungulates, primates and canids there was a common selective force favouring the development of spontaneous ovulation. The number of breeding females often outnumbers the number of breeding males and the latter usually compete for this privilege (see p. 37). Temporary pair bonding is common between the female and male at the time the female comes into oestrus, and to ensure conception in all females Conaway argues that it would be a great advantage to spread the oestrus periods of females randomly over a period of time. Synchronization of oestrus, which commonly occurs in induced ovulators, would be disadvantageous. From the opposite point of view, Weir and Rowlands (1973) argue that induced ovulation would free the female from cyclic behaviour and allow mating as soon as the male was available and/or fecund and that this would be advantageous for solitary species such as the cat, but not for gregarious species such as the rabbit and some voles. Daly and Wilson (1978) follow Weir and Rowlands in suggesting the relative advantages of induced ovulation to solitary species, particularly those living at low densities.

The possible evolutionary progressions from spontaneous to induced mechanisms in the follicular and luteal phases allow the reproductive process to become more efficient in the recognition of true pregnancy and the return to the oestrous state in spontaneous ovulators (Types I and III) or to become completely free from the cyclic nature of oestrus (Types II and IV) and only dependent upon a good 'quality' mating stimulus for proper activation of the corpora lutea. Further streamlining of the follicular and luteal phases occurs by shortening them when they are spontaneous. Only the higher primates keep the relatively long follicular phase and in the Type I and II species the pseudopregnant, active, luteal phase is rarely as long as pregnancy (as in the ferret and domestic dog) and usually substantially shorter, as in the domestic mammals and other examples of these groups (see Table 2.1). If, as seems likely, Weir and Rowlands are right in their assumption that spontaneous ovulation is the primitive condition, then the evolution of induced ovulation in the solitary or rapidly reproducing species appears to be the most acceptable argument given our present knowledge of its occurrence.

Courtship and mating

Females in oestrus are attractive to males, as without a male the oestrus will be wasted and no fertilization can occur. The attraction will usually be by visual or olfactory means and copulation may be almost immediate with very little courtship as in many rodents, or extend to involve an elaborate behavioural pattern which may last for several days, as in the orang-utan (Mackinnon, 1974a), where a long consortship preceeds mating. The details of courtship and mating have been the subject of many reviews (Dewsbury (1972) for males, Beach (1976) for females, Kleiman and Eisenberg (1973) for canids and felids. Eltringham (1979) for African herbivores and carnivores and Chalmers (1979) for primates). Therefore the emphasis here is on the ecological aspects of mammalian mating behaviour, elucidating possible relationships between the ecology of mammals and their mating systems and discussing the behavioural and physiological factors which enhance an individual's success in mating and in producing members of the next generation.

Mating systems are important reproductive variables because they determine whether an individual will have a relatively constant success in reproducing or whether he (or she) will perhaps have a good or poor chance of reproducing successfully. Reproductive success, or *fitness*, in an individual is defined as the success with which an individual is able to pass on its genetic make-up to the next and successive generations. However many females a male might be assumed to control, or however many matings a male might perform, the ultimate test of his reproductive success can be measured purely in genetic terms. Female mammals have a very different strategy (adaptive syndrome) for success in reproduction than males in that they devote much time and energy to pregnancy, lactation and maternal care. Males only need to fertilize an ovum to benefit from this female strategy and thus normal intraspecific competition will include competition to mate with a female or females. Females are not able to produce offspring above a certain limit of numbers and so females are a resource over which the males compete. Under these conditions it would seem to be advantageous to a male to enter into a mating system where he had the opportunity of mating with more than one female so a particularly interesting question to ask is 'under what circumstances does a male limit himself to mating with one female?' This is one of the questions that comparative studies, often of related groups of species, are able to answer with some degree of satisfaction. Much of the early work on the evolution of mating systems was carried out with birds particularly in mind (Verner and Willson, 1966; Orians, 1969), but later authors have looked at the evolutionary and ecological aspects of mating systems in mammals or groups of mammals (Geist, 1974; Jarman, 1974; Crook *et al.*, 1976; Kleiman, 1977; Ralls, 1977; Jewell, 1977; Clutton-Brock and Harvey, 1979) or at least considered birds with mammals in mind (Emlen and Oring, 1977).

A mating system (Jewell, 1977) comprises those social relationships which permit individuals to find and compete for a mate; possibly to establish long-lasting relationships; to induce a potential mate to copulate; and to provide parental care for the offspring. These relationships may be broadly categorized into monogamy, polygamy and promiscuity. Monogamy is the case where the bond between the sexes is between one male and one female. Polygamy may be one of two types, polygyny being where one male has bonds with several females and polyandry where one female has bonds with several males (not known in mammals). Promiscuity occurs where there are no bonds between the sexes, and is really a type of polygyny. There are difficulties in using this classification in that promiscuity does not necessarily mean that there is no selection of mates, and without further classification, such as successive polygamy and simultaneous polygamy (Daly and Wilson, 1978), some different mating systems might be confused. Jenni (1974) discusses the uses of these general terms and Selander (1972) and Ralls (1977) offer alternatives. As promiscuity has the same genetic effect as polygamy they are discussed together.

Monogamy or polygamy

The particular mating system associated with a species of mammal is relatively easy to define in terms of the broad classification outlined above, but the ecological and evolutionary pressures leading to the particular system are not as simple to understand. An outstanding piece of work covering all the orders of mammals by Kleiman (1977) has attempted to answer these questions for monogamous species.

Monogamy

It is convenient to start with examples of this relatively simple mating system involving just two individuals at any one time, and then look at some of the evolutionary pressures which might have led to it.

Monogamy occurs in only a small minority of mammals and although many species have not been studied in enough detail the figure of less than 3 per cent of mammalian species having been reported as monogamous (Kleiman, 1977) is probably not far from the truth. It is the most common form of social organization in the dogs and foxes (Canidae) and in primates it is found in the marmosets and tamarins (Callitrichidae), gibbons (Hylobatinae) and indris (Indriidae). It also occurs in some families of marsupials, insectivores, bats, seals, rodents, baleen whales, and artiodactyls as well as in other families of carnivores and primates.

A familiar example of a monogamous species is the fox (*Vulpes vulpes*) where the male and female rear the family of cubs together and the male helps the female to rear the young (Lloyd, 1977). There is a possibility that sometimes two vixens may be present but the fox and many other canids are generally taken to the monogamous (Kleiman and Eisenberg, 1977). In the wolf (*Canis lupus*) the young stay with their parents and mature in their second year, forming a family group. A similar monogamous mating system is seen in the beaver (*Castor fiber*) (Wilsson, 1971), where the male remains with the female and her offspring for more than one season and the young show delayed maturity; the beaver matures later than any other rodent, at two years of age.

Less familiar species showing a simple type of monogamy are the elephant-shrews *Rhynchocyon chrysopygus* and *Elephantulus rufescens* in which the adult pairs have completely overlapping territories but are not often observed together. They are reported as never occurring in family groups (Rathbun, 1976; Kleiman 1977). Similarly adult Kirk's dik-diks (*Madoqua kirki*), small bovids, are rarely seen together and the young are driven off shortly before a subsequent birth (Hendrichs and Hendrichs, 1971; Hendrichs, 1975).

Kleiman (1977) has defined the latter, less familiar type of monogamy as Type I (facultative) monogamy and contrasts this with the former, more commonly observed family group of an adult pair and more than one generation of young which she calls Type II (obligate) monogamy. Other differences between the two types of monogamy are listed below but Kleiman points out that not all monogamous species are easily placed in one type or the other. Monogamy seems to be related to a low reproductive rate but comparable non-monogamous species in the primates and artiodactyls have similar, low, reproductive rates. Sexual maturation may occur later in some monogamous species than in related non-monogamous species and the age at puberty appears to be later in Type II species than in Type I species. In Type II species the young often help in the care of the next litter and so create a family unit (e.g. in marmosets, tamarins, gibbons, jackals, dwarf mongooses and beavers).

Kleiman comes to the conclusion that Type I monogamy, where family groups are rarely seen together, may have developed in response to a low population density where only a single female may be available to a male for reproduction. Such a situation might arise if food resources were concentrated into rich but clumped patches. Type II monogamy may have evolved in response to (a) a situation where more than one individual is needed to rear the young or (b) a situation where the carrying capacity of the habitat is not great enough to allow another female to rear a litter in the same home range, or both (a) and (b).

In a study of primate ecology and social organization Clutton-Brock and Harvey (1977) come to similar conclusions over the possible reasons for the evolution of monogamy in this order. There are at least 13 species of primate which are monogamous (four tamarins and marmosets, all six observed gibbons and siamangs, two titis (New World species of *Callicebus*) and a lemur). They follow Trivers (1972) in suggesting that monogamy is likely to occur where the male leaves more offspring by assisting the female to rear their young. In primates male assistance takes the form of carrying young and territorial defence, and monogamy is most likely to evolve where male assistance is most necessary (Brown, 1975). In the tamarins and marmosets the usual occurrence of twins probably explains their monogamy (Clutton-Brock and Harvey, 1977). Clutton-Brock and Harvey argue that monogamy might be expected where territory size is limited by the area which can be effectively defended by sight and sound, and where the area is not large enough to support polygyny. In this territorial situation monogamy might be seen in species living at relatively low population densities because of large body size, specialized food habits or the high intensity of interspecific competition. All the primates in which monogamy occurs are territorial and live in habitats where intraspecific competition is consistent; most occur at relatively low population densities as a result of large body size (indri, lemur, siamangs) or low food supplies (one of the *Callicebus* species), but there are exceptions. Although the possible reasons for the evolution of primate monogamy are not exactly like those proposed for mammals in general by Kleiman (1977) they are basically the same. Without more detailed information about the food supply and social behaviour of each species they cannot be compared further, except to say that Kleiman places most of them in the Type II category.

Polygamy

Polygamy in mammals occurs as polyandry, polygyny and promiscuity as explained on page 34. An example of a polygynous species is the red deer (*Cervus elaphus*) where the stag controls a rutting area and establishes a harem on it; unsuccessful stags are chased off and have little opportunity to mate (Lowe, 1977). Another is the common zebra (*Equus burchelli*) which has a small group of females associated with a single male all the year round (Klingel, 1968, 1969). Many pinnipeds are also polygynous, for example the northern fur seal (*Callorhinus ursinus*), which has a harem of about 40 females to one bull (Bartholomew and Hoel, 1953), and where the successful male should produce about 80 male progeny, whereas a male that never establishes a territory among the females in a rookery would leave few or no progeny (Bartholmew, 1970). Examples of promiscuity are seen in tree squirrels, some bats, elephants, and giraffe (*Giraffa camelopardalis*) (Crook *et al.*, 1976, Jewell, 1977). In these cases the male does not have a long association with the female, although, as in the case of some bats, they may belong to the same group. Polygyny, and often promiscuity, is associated with sexual dimorphism in size with males being larger than females. This is partly, if not totally (see Ralls, 1977), the result of sexual selection (Darwin, 1871) where individuals of one sex (usually males) compete with each other for the opportunity to mate with a female or females (intrasexual selection; Huxley, 1938) and/or where members of one sex, usually females, choose mating partners (intersexual or epigamic selection; Huxley, 1938). Darwin separated sexual selection from natural selection but they are really both the same thing as the end result is the survival of the individual (or genetic complement) with the greatest reproductive success (Campbell, 1972).

Based on the premise of sexual selection and various ecological parameters Orians (1969) suggested a number of reasons for the evolution of monogamy and polygamy, particularly for passerine birds; the ideas are developed from a paper on passerine birds by Verner and Willson. (1966). Orians argues that since polygyny is very advantageous to the male the presence or absence of polygyny must be determined by the advantage or disadvantage to the female. Female choice will be based upon (a) the genetic quality of the male and (b) the quality of the habitat on which the male is present because in many cases (of birds) the male is involved in providing food for the young and/or protecting young from predators. However, a female in a good quality territory may have more females in optimal habitats as co-wives and therefore the average reproductive success would decline; this would be because there would be fewer resources per individual, a high density of predators and the male would be shared between more females. Thus polygyny will evolve only when the quality of the habitat occupied by mated and unmated males is sufficiently different to make mating with a bigamous male advantageous for the female. In other words the variability of the habitat will reach a *polygyny threshold* (Fig. 2.2). Under these circumstances one would expect similar quality habitats to have no polygyny and similar reproductive success between individuals. Orians emphasizes a number of factors which will affect the fitness of a monogamous or polygynous species. Some of the major points are picked out below. If there is no parental care from the male there will be

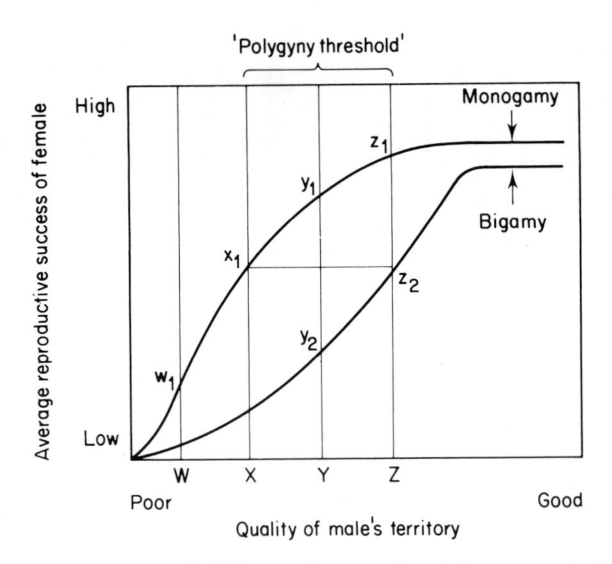

Fig. 2.2 The polygyny threshold. Female reproductive success is determined by the quality of the male's territory. Thus a female with a monogamous male on a territory of quality X will have the same reproductive success as a female with a bigamous male on a territory of quality Z. Redrawn after Verner and Willson (1966), Orians (1969) and Daly and Wilson (1978).

no difference in the fitness of the females and polygyny would be prevalent; the choice is based solely on the genotype of the male unless male competition is the deciding factor. Also, polygyny should be most prevalent in habitats such as early successional ones where the variation in food supply is considerable.

On the basis of these and other bird-orientated predictions Orians stated that polyandry should be rare (there are only a few examples in birds and none in mammals) and that monogamy should be relatively rare in mammals. These predictions seem generally applicable at first sight, but Ralls (1977) has drawn attention to some aspects which are really limited in their generality to some bird groups (rather than all birds and mammals). She points out that most mammals do not meet the conditions on which female choice is based. Not all females choose to mate with particular males (e.g Hamadryas baboons, *Papio hamadryas*, and polygynous pinnipeds). Also, the female does not necessarily raise her young on the territory of the male and depend only on its resources, especially in many highly polygynous species. The main force opposing polygyny is the need for parental care, according to Orians (1969) and this is often not the case; it might account for Type I monogamy but certainly not for Type II. An adequate general model will have to include factors other than a large male parental investment to oppose the evolution of polygyny. However, Orians' model may have relevance to mammal species where females do make a choice of a particular male.

At this point it is best to accept that the reasons for the evolution of a polygynous mating system are not clear-cut. There are at least three other models and none has general applicability (Altmann *et al.*, 1977). They are:
(a) that it is a by-product of an unbalanced sex-ratio (Mayr, 1939; Wittenberger, 1976) which is hardly a general explanation;
(b) that it enables males to regulate their population density because they are fully informed about their own reproductive activity and can thus cease mating when local density becomes too high (Wynne-Edwards, 1962). The fact that when females (birds) appear on breeding areas they are able to obtain males and raise young (Orians, 1969) seems to be contrary to the Wynne-Edwards idea;
(c) that it is the result of fighting and display among males for the exclusive possession of females (Darwin, 1871). The harem systems of many pinnipeds, cervids and bovids, and Hamadryas baboons fit this explanation because, as mentioned above, the females appear not to choose their mates and the males herd the females and compete among themselves in order to maintain exclusive access to the female group (Bartholomew, 1970; Jarman, 1974; Kummer, 1968). However, in other polygynous mammals, for example Uganda kob (*Kobus (Adenota) kob*) territorial males are chosen by the female (Buechner, 1974) and female fallow deer (*Dama dama*) congregate around territorial rutting bucks (Chapman and Chapman, 1975).

An insight into the ecological conditions associated with particular mating systems is provided by Emlen and Oring (1977) who divide the types of polygyny to good effect. Emlen and Oring (1977) consider two variables, the spatial distribution of resources and the temporal distribution of mates, and the ways in which they might interact to favour polygyny. If important resources are clumped it is possible for a small percentage of the population to monopolize a large proportion of the available resources. Under these circumstances sexual selection and the variance in reproductive success are likely to be high and the environment thus has a high polygyny potential (see discussion of polygyny threshold, p. 37). The benefits derived from resource defence are of no use if the females are highly synchronized in their oestrous behaviour. Polygyny is advantageous only if the females are asynchronous and the time taken to service a single partner is insignificant compared with the total time that females are receptive and able to be fertilized. Otherwise other males will inseminate other females while the first male is courting and mating only one female. These two variables are related to each other to produce ecological conditions which favour polygyny as the resources become more clumped and the availability of mates becomes less synchronous (Fig. 2.3).

Emlen and Oring (1977) divide polygyny in its broadest sense (i.e. where males control access or gain access to many females) into three main types.

(1) *Resource defense polygyny* where males control access to females indirectly by monopolizing critical resources. The female choice of mate should be influenced by the quality of the defending male and by the resources under his control (in a territory). An example of this type of polygamy is shown by the yellow-bellied marmot (*Marmota flaviventris*) where

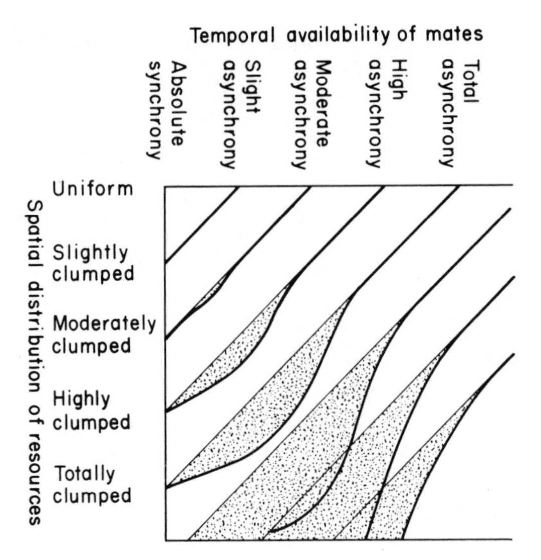

Fig. 2.3 Diagram illustrating the possible potential of the environment for polygamy (indicated by the perpendicular height of the shaded area) and its relation to the distribution in space of resources and the availability in time of receptive mates. After Emlen and Oring (1977). Copyright 1977 by the AAAS.

suitable habitat is severely restricted in a high mountain environment; the best habitats are rocky outcrops which provide cover from predators and sites for hibernation and these comprise male territories which attract females (Armitage, 1974; Downhower and Armitage, 1971; Andersen *et al.*,

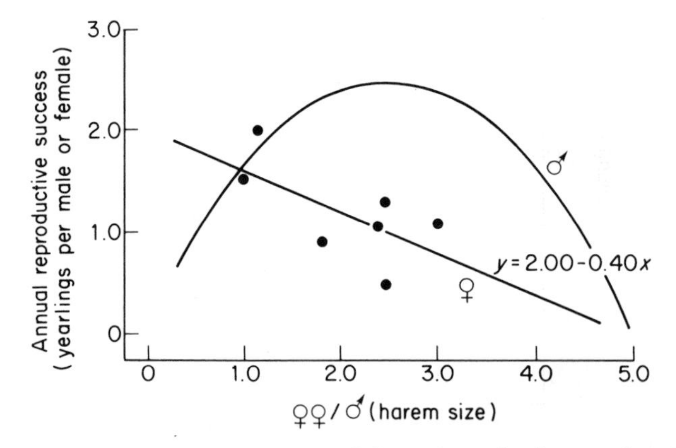

Fig. 2.4 The number of yearlings per adult marmot for harem females and the resident male as harem size increases. The points show the number of yearlings per female (annual reproductive success) and the average number of females per harem. The linear regression is significant ($0.05 > P > 0.01$, $t = 2.64$). For explanation of the model 'fitness' curve labelled with the male symbol see text. Redrawn from Downhower and Armitage (1971).

1976). The localities are vigorously defended by the territorial males and this polygynous system permits maximal numbers of females to be associated with a minimum number of males, thus maximizing the potential production of young. The territories are relatively stable, with a mean length of residency of the colonial male of 2–4 years, so that there is an optimum social environment for reproduction with, usually, little disruption. Downhower and Armitage (1971) proposed a model for polygyny in this species in which the male maximizes his fitness by having a colony of 2–3 females (Fig. 2.4). They noted that the yearly reproductive success of females declined with harem size (Fig. 2.5) and so a female might be better off if she was monogamous; thus the mating system seems to be a compromise between the male and the female strategy for reproductive success. However, this may not be the case if, as Elliott (1975) suggests, the yearly reproductive success is not a good measure of lifetime reproductive success; it may be that some females suffer from low reproductive success when they are subordinate but gain over the monogamous state when they live to be dominant. Armitage (1965) presents evidence suggesting that this is so.

(2) *Female (or harem) defense polygyny* where males control access to females directly, usually by virtue of female gregariousness which is not related to reproduction. During mating many ungulates such as the African impala (*Aepyceros melampus*) and waterbuck (*Kobus defassa*) divide the preferred habitat into male territories and aggregations of females and

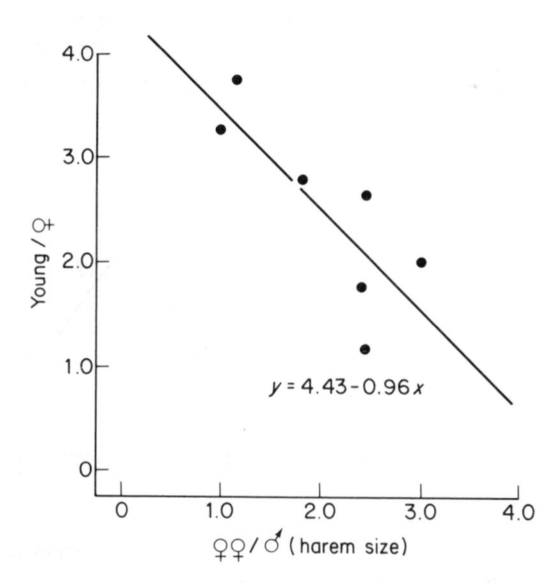

Fig. 2.5 The number of young per female marmot as harem size increases. The points show the number of young per female and the average harem size for seven harems. The regression is significant ($0.05 > P > 0.01$, $t = 3.06$). Redrawn from Downhower and Armitage (1971).

young move into these territories for cover and food. The polygyny seen here may verge on promiscuity as the females are not herded by aggressively excluding all other males. Male pinnipeds show the same type of polygyny by taking advantage of the gregarious hauling-out behaviour of the females which are about to give birth. In some seal species (e.g. elephant seals) the females are directly defended and males which try to encroach are chased off by the dominant male (Bartholomew, 1952).

(3) *Male dominance polygyny* where males do not directly defend females or resources essential to females, but rather sort out among themselves their relative positions of dominance. Females choose males primarily on the basis of male status. Examples of this type of polygamy are seen in the promiscuous mating system of the Soay sheep (*Ovis aries*), the buffalo (*Syncerus caffer*) (Grubb and Jewell, 1973; Jewell, 1977), and many bat and rodent species. Alternatively a 'lek' mating system is seen where the females are relatively asynchronous in their periods of sexual receptivity to males and the latter generally remain active for the duration of the population's breeding season. The intensity of male–male competition together with the long mating period leads to a stable aggregation of advertising males. The 'lek' is a communal display ground where males congregate simply to attract and court females and to which females come for mating (Wilson, 1975). It is common for the central positions to be taken up by older, more dominant, males which achieve greater breeding success. Jewell (1977) reviews the lek in Uganda kob and states that it is occasionally seen in the topi (*Damaliscus korrigum*). Bradbury (1977) describes lek mating behaviour in the hammer-headed fruit bat (*Hypsignathus monstrosus*) found in the African lowland rain forest of Gabon. In this latter species the lek sites, or traditional mating areas, are spaced about 14 km apart and each one consists of between 20 and 135 males. They are typically situated along the borders of rivers where each displaying male defends his territory of 10 m diameter. The males display from dusk to midnight and again before dawn by flapping their wings and emitting a loud monotonous honking. Females visit the leks, examine a succession of males and eventually select one for mating; only a few males do most of the mating.

Thus it is possible to see many variations in the mating systems of mammals by looking at different species and also by looking within species. Polygyny is particularly advantageous to some males, and studies of mating systems usually assess the degree of sexual selection which follows.

Behavioural and physiological factors affecting reproductive success

Behavioural factors and fitness

Behavioural and physiological factors, as well as morphological factors, all play a part in increasing the reproductive fitness of a male and it is these factors, together with an assessment of whether preferential mating really does lead to increased fertilization and greater reproductive success, which will be discussed next.

We have already seen that in some mating systems the female may choose the male with whom she will mate and that this is called epigamic selection. Ralls (1977) suggests that the importance of epigamic selection may have been overlooked in the past. However, there is little doubt that in many polygynous species the males have to compete for the opportunity to mate and thus show strong intrasexual selection. The interaction of these two selective processes will determine how successful a male is in reproduction, assuming that no problems lie in the path of the mated or fertilized female.

A good example of behaviour affecting reproductive success as measured by mating is that of the elephant seal (*Mirounga angustirostris*) (Le Boeuf and Petersen, 1969; Le Boeuf, 1972, 1974). In this seal the males compete for the possession of a number of females by fighting in a violent manner in order to establish a dominance hierarchy. Only the dominant males mate extensively and only 4% of the males achieve 85% of all matings. Since reproductive success must really be taken over the whole lifetime of the male it is interesting to note that in this case the seasonal picture does seem to reflect the lifetime reality of very great variation in the reproductive success of males. Le Boeuf (1974) was able to observe that

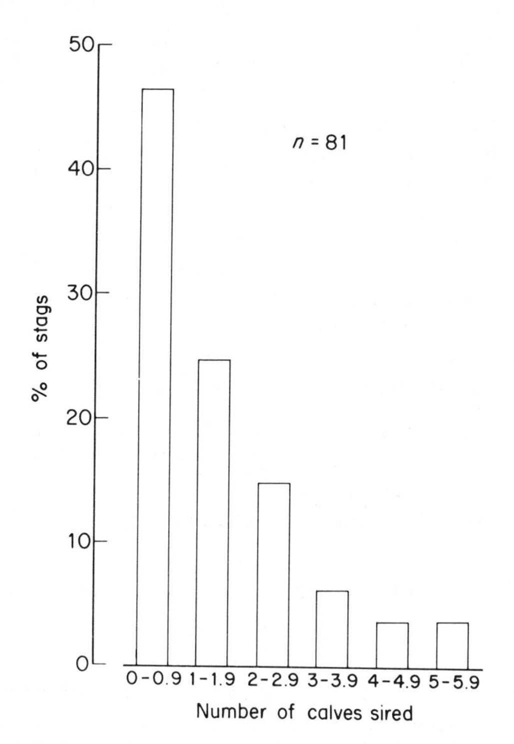

Fig. 2.6 The distribution of reproductive success (% of stags siring various numbers of calves) among adult red deer stags over 4 years old within seasons. Redrawn from Gibson and Guiness (1980).

only about 10% of the males survive from birth to the age of 5 or 6 years when they join the breeding population and past 6 years of age the chance of a male surviving to compete the next year is about 50% per year. The dominant males were estimated to be siring about 50 young per year and one long-lived individual probably sired 200 over a four-year period.

Another long-term study where behaviour and reproductive success have been measured is that of Gibson and Guinness (1980), who studied male reproductive success in the red deer (*Cervus elaphus*) over five rutting seasons. Most older stags attempt to breed by fighting for the possession of the older hind groups (harems) but sustained rutting activity rarely appears before the attainment of adult body weight at about 5 years of age (Lincoln and Guinness, 1973). The number of calves sired per season by the stags was estimated from the long-term data by multiplying the number of hinds in a stag's harem on each day of the rut (a few weeks) by the proportion of hinds conceiving on that day and then summing these values for all the days of the rut. The proportion of hinds conceiving was calculated by multiplying fertility (the proportion of hinds conceiving per year) by the relative proportion of conception on each day of the rut, worked out by looking at the combined distribution of births for all the years. These estimates of reproductive success in the adult stags (over 4 years old) indicate that there were large differences in individual success within seasons (Fig. 2.6). The variation in reproductive success among adults was not primarily age-dependent (Fig. 2.7) and they came to the conclusion that the magnitude of consistent individual differences in reproductive success over successive seasons would make the equalization of lifetime success by differential mortality improbable; there was an eleven-fold difference between the average success per season of the most and least successful individuals. Thus the argument of Geist (1971), that if breeding incurs energetic costs which decrease survival, males which are most successful within a short

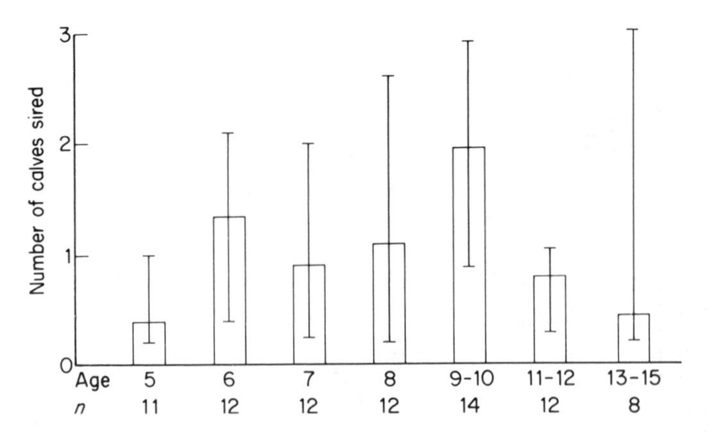

Fig. 2.7 Reproductive success (number of calves sired per season) in relation to age in adult red deer stags. The median (columns) and interquartile range (vertical bars) for each age class are shown. Redrawn from Gibson and Guiness (1980).

Table 2.2 Production of young by rabbit does of different social status living within an 82 acre paddock. After Mykytowycz and Fullagar (1973).

Breeding season	Social status*	No. of potential breeders	No. of actual breeders	No. of litters produced	No. of young surfaced	Mean young production per doe	No. of young surviving till next breeding season	Mean survivial of young per doe
1967	1	5	5	12	66	13.2	19	3.8
	2	7	6	8	38	5.4	8	1.1
	3	2	1	1	7	3.5	1	0.5
1968	1	8	8	24	133	16.6	63	7.8
	2	14	14	27	137	9.7	63	4.5
	3	5	3	5	23	4.6	9	1.8
1969	1	20	12	13	53	2.6	0	–
	2	36	9	9	35	1.0	0	–
	3	30	0	0	0	0.0	0	–
	4	11	1	1	3	0.3	0	–
1970	1	2	2	11	58	29.0	24	12.0
	2	3	3	11	44	14.6	21	7.0
	3	1	1	3	14	14.0	12	12.0
1971	1	8	7	8	44	5.5	0	–
	2	15	8	8	38	2.5	0	–
	3	3	2	2	8	2.6	0	–
	4	2	0	0	0	0.0	0	–

*1 = dominant, 2 = sometimes dominant, 3 = subordinate but interacted with other females, 4 = subordinate and did not interact with other females period might not achieve higher lifetime success because they breed in fewer years, appears to be improbable.

Short term studies showing differential breeding success in the male have been carried out on primates (Bernstein, 1976), ungulates (Owen-Smith, 1977), bats (Bradbury, 1977) and rodents (Farentinos, 1972; Thompson, 1977). These suffer from the argument put forward by Geist (1971) and detailed above, as well as the possibility that differences in reproductive success over a short time period could simply represent age-dependent variation (Rowell, 1974). However, a more devastating criticism has come from Ralls (1977) who suggests, and gives evidence that, field measures of male reproductive success could be poorly correlated with fertilization success. Oestrus does not always coincide with ovulation in rhesus monkeys (Loy, 1970) and Duvall *et al.*, (1976) showed that in the rhesus monkey (*Macaca mulatta*) physiological evidence of paternity indicated that the dominant male could not have fathered more than 7 out of 29 infants born, which was not significantly different from the number expected by chance.

Despite the possible exceptions and the criticisms, it does seem very likely that behavioural (dominance) relationships have a major part to play in the reproductive success of the male. Sexual selection might even be intensified by the female if she will only allow copulation to be completed by a dominant individual. This is indeed the case in elephant seals where the female protests loudly as the male mounts and attracts the attention of the other males (Cox and LeBoeuf, 1977). The other males respond by attempting to dislodge the mounting male and only if he is dominant will he succeed in copulation. Thus the female makes it more difficult for young, subordinate males to copulate.

It has already been suggested that female reproductive success might be affected by dominance/subordinance relationships in the yellow-bellied marmot. This effect is clearly present in rabbit (*Oryctolagus cuniculus*) populations in Australia (Mykytowycz and Fullager, 1973) (Table 2.2). In these enclosed populations of rabbits in two years out of three the mean number of young surviving per doe is directly related to the social status of the doe, the dominant does each producing between four and eight times the number of young surviving per doe to the next breeding season as the subordinate does. The dominant does produced 51% of the young of known parentage whereas the second-ranking does produced 42% and the subordinates 7%. The effects of the dominance were to delay the start of breeding of the subordinate does and to allow the good survival of the young of the dominant does; mortality of young increases through the breeding season in the climatic conditions of south-eastern Australia.

Thus, although there are some exceptions, it seems that behavioural dominance can have a fundamental effect on the reproductive success of male and female mammals.

Wilson (1975) suggests some other behavioural mechanisms for improving the reproductive success of males. One of these is the guarding of the female after mating; male elephants keep company with an oestrous female and copulate several times if they are not prevented from doing so by other males (Eltringham, 1979). This form of guarding presumably could happen in many polygynous mating systems but not in highly promiscuous ones. Wilson's other factors helping to improve an individual male's reproductive success after female copulation with another male include induced pregnancy failure and infanticide. Induced pregnancy failure is meant as a general term covering the induced failure of implantation as well as the resorption of embryos. The first phenomenom occurs in laboratory colonies of species of rodent such as the deer mouse (*Peromyscus maniculatus*) (Eleftheriou *et al.*, 1962) and house mouse (*Mus musculus*) (Bruce, 1959) and also in the lemming (*Dicrostonyx groenlandicus*) (Mallory and Brooks, 1978). The second phenomenon of termination of pregnancy after implantation occurs in the prairie vole (*Microtus ochrogaster*) (see Chapter 3). Suffice it to say that the presence of a 'strange' male (i.e. a male with which the female has not previously mated) near a pregnant female often leads to pregnancy failure and a return to oestrus. A similar result is achieved by a more gruesome means in infanticide, which has been shown to occur in lemmings (Mallory and Brooks, 1978), lions (*Panthera leo*) (Bertram 1975, 1976) and in the Indian langur (*Presbytis entellus*) (Hrdy, 1974). In these situations the male killing the young advances the time of the next oestrus of the female, improving his chances of mating and therefore his fitness. It seems a waste for the female if males kill young like this, but the males are probably not related to the females so that they are not killing any of their own progeny or others closely related to them. It may even improve the reproductive fitness of the female if in lions, which have communal suckling, all females give birth at the same time and so peer groups of young males grow up together and have a better chance of survival than when brought up singly.

Physiological factors and fitness

Before the mated female can be fertilized the sperm from the male must usually move from the place of ejaculation (vagina or uterus) up to the oviduct where it meets the ovum coming down. If more than one male has mated the female there is the possibility of multiple paternity in a polytocous species, as has been shown by electrophoretic techniques in the deer mouse (*Peromyscus maniculatus*) (Birdsall and Nash, 1973). The sperm do not actively compete except perhaps in terms of population number (Parker, 1970), so that males which produce large numbers of sperm are at a selective advantage. It seems that the main reduction in the number of sperm is carried out by the successive barriers of the cervix, uterus and uterotubal junction (in the rabbit; Cohen and McNaughton, 1974) and that particular sperm are selected for their fertilizing ability as sperm from the oviduct fertilizes another female, when artificially inseminated, to a greater extent than sperm from the uterus. Thus it may be possible for sperm to compete in counteracting these (unknown) selective mechanisms.

Prevention of further copulation is a further way of safeguarding fertilization from the original mating and if this is not done behaviourally it may be done physiologically by a 'copulation plug'. The copulation plug is formed from secretions of the male reproductive accessory glands (see Chapter 1) and has been observed to prevent successful intromission in at least the bank vole (*Clethrionomys glareolus*) on some occasions (Milligan, 1979). Such plugs may hinder, if not prevent, further successful mating until they drop out; copulation plugs are found in many species of rodents and some bats, insectivores and marsupials.

Pregnancy

The time from fertilization until parturition is usually regarded as being a species' characteristic. If the length of the period of relatively fast growth to parturition is plotted against the cube root of birth weight (or birth length), then species in the same order or similar taxonomic group seem to possess very similar relationships (i.e. similar growth rates) (Hugget and Widdas, 1951; Frazer and Huggett, 1973).

Thus, although there is much variation in foetal growth rate, certain orders may be characterized as having relatively fast or slow foetal growth rates. Those groups which show relatively slow growth rates are the hystricomorph rodents, the bats and the primates. This variation in foetal growth rate is not always easy to explain in terms of adaptation to the external environment (Weir and Rowlands, 1973). Slow growth may be related to poor feeding conditions for the mother; Weir and Rowlands (1973) suggest that this might have been the case during the evolution of the hystricomorph rodents. However, it is possible that post-natal influences are important; McKeown *et al.* (1976) suggest that if a species has delayed physical and sexual maturity then it is likely to have evolved, at the same time, slower prenatal growth alongside the retarded postnatal growth.

Litter size, the time to weaning and the maturity of young at birth may be considered in relation to ecological as well as physiological variables and are therefore discussed in this chapter in the section on parturition and weaning as well as in relation to life-history variables in Chapter 4. Foetal growth, however, shows some variation between very similar species as it may be subject to delays at the start and end of pregnancy as well at the time of implantation; this is discussed below. Some of these physiological variations may be interpreted as being adaptive and also, sometimes, having direct links with environmental conditions.

Delayed fertilization

The phenomenon of delayed fertilization has developed in species of vespertilionid and rhinolophid bats in connection with the habit of hibernation and is closely linked with delayed implantation (see next section).

Oxberry (1979) reviews both delayed fertilization and delayed implantation in bats. He points out that as hibernation has evolved as a means of energy conservation in the stressful situation of low ambient temperatures, it can therefore be expected to influence, or be influenced by, reproduction which is also energetically expensive. Delayed fertilization is characteristic of many North American verspertilionids, and most hibernating

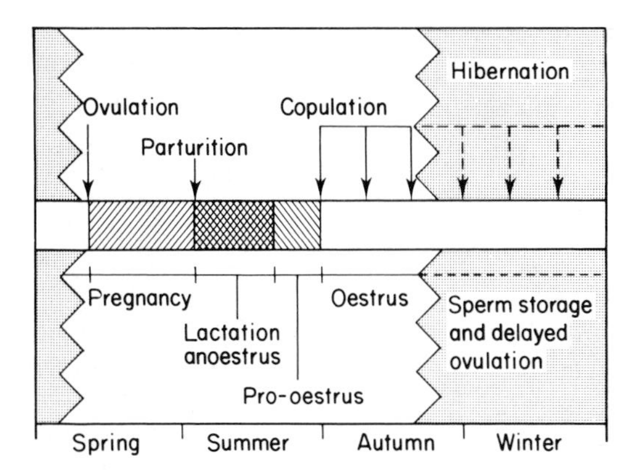

Fig. 2.8 The pattern of events in the reproductive biology of bats with delayed fertilization. Redrawn from Oxberry (1979).

vespertilionids and rhinolophids in temperate Europe, Asia and Australia (Fig. 2.8). There is a brief pro-oestrus period in late summer and oestrus and subsequent copulations are initiated in late summer and early autumn, after which they usually enter hibernation. Spermatozoa are stored in the uterus until after permanent arousal in spring, when ovulation, fertilization and gestation take place. Parturition occurs in late spring or early summer. Hibernation seems to temporarily arrest the normal events of reproduction. Specialized, large mature Graafian follicles remain in the ovary throughout hibernation and not only are they unusual in having a large, hypertrophied, discus proligerus but this area is also rich in glycogen for the nourishment of the ovum. Wimsatt (1960) suggested that the delay in ovulation involved the interplay of at least three factors. Hibernation itself would depress cellular metabolism and reactivity; pituitary gonadotrophic function might be dissociated from its neural regulation; and the neural regulation of pituitary gonadotrophins might be triggered by environmental and/or internal stimuli. Oxberry concludes that there is insufficient pituitary LH present for ovulation to take palce, especially during the first half of hibernation. He believes that this may be the result of the lowered metabolic rate in hibernation and/or of the neural regulation of the pituitary, triggered by environmental stimuli.

It has been shown conclusively that fertilization in the spring may be accomplished by the stored sperm (Gustafson, 1979; Racey, 1979). Sperm from more recent copulations performed during the temporary arousal of the bat from hibernation, or after hibernation, may also affect fertilization. The fact that fertilization is possible in the spring, after spring copulation, stimulates the question 'why bother to delay ovulation and store sperm?'. There is no really satisfactory answer to this except to repeat that the reproductive events in these hibernating bats are almost certainly linked with the physiology of torpor. Thus there are plausible theories concerning the presence of delayed fertilization in bats but no really convincing arguments to explain it.

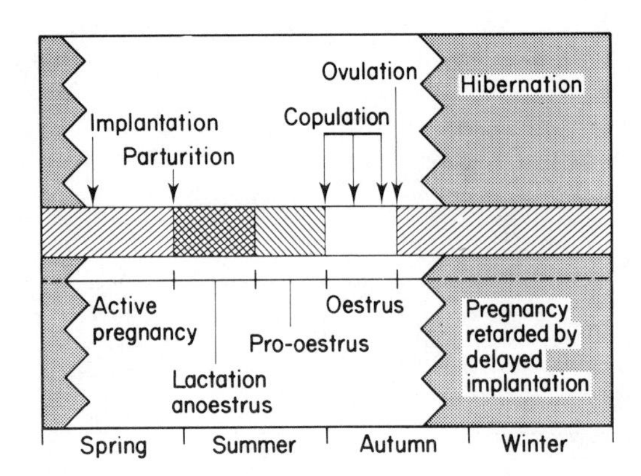

Fig. 2.9 The pattern of events in the reproductive biology of bats with delayed implantation. Redrawn from Oxberry (1979).

Delayed implantation

It is convenient to remain with bats in the preliminary discussion of delayed implantation as this is the second method of prolonging pregnancy to be seen in this group. In hibernating bats delayed implantation occurs in the Old World vespertilionid genus *Miniopterus* and in the rhinolophid genus *Rhinolophus* (Oxberry, 1979). The sequence of events is shown in Fig. 2.9. A brief pro-oestrus period occurs in late summer and oestrus and copulation follow in the autumn. Ovulation occurs immediately so that fertilization and initial development of the blastocyst occur. The bats enter hibernation with unimplanted blastocysts in their uteri and only in the spring, after arousal, does the blastocyst implant and embryonic development resume. This is a further adaptation of reproductive physiology to cope with the constraints of hibernation, namely low metabolism, and reduce the energy required to overwinter in a pregnant condition. In non-hibernating bats only the fruit bat (*Eidolon helvum*) has a delay, apparently related to the dry season in an equatorial and tropical environment (Mutere, 1967).

Delayed implantation occurs in many other species of mammal in unrelated orders (Renfree and Calaby, 1981). Figure 2.10 shows some of these patterns of reproduction incorporating delayed implantation and stresses the point that it occurs most commonly in eutherians in the seals and carnivores. The figure shows those species which have obligate delayed implantation (Wimsatt, 1975) when delay in the implantation of the blastocyst is a genetically fixed component of every pregnancy. However, a facultative delay also occurs commonly in some rodents (e.g. mice and voles) where the blastocyst undergoes a quiescent period under stressful conditions (usually lactation of a previous litter) but this is not so in every pregnancy. These definitions, as Wimsatt (1975) points out, mean that although the lactational delay of many Australian marsupials displays some facultative components and allows reproduction to be very well

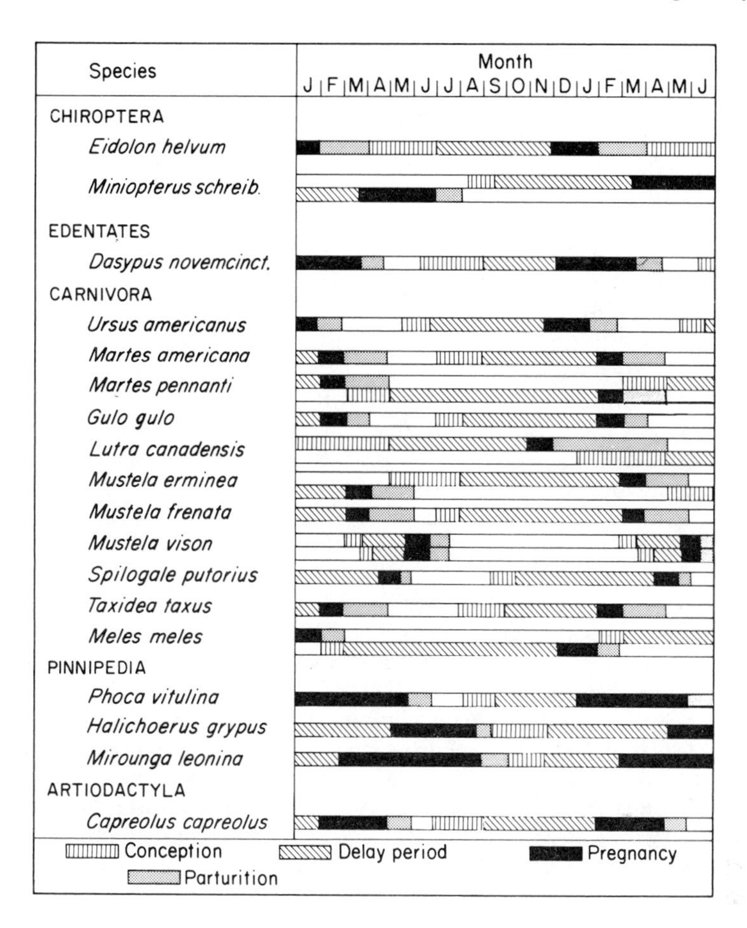

Fig. 2.10 Patterns of delayed implantation in some mammals. Modified from Wimsatt (1975).

adapted to the specific environmental and lactational conditions, it is still of the obligate type because it is genetically fixed in every pregnancy. The nature of delayed implantation in marsupials is fully described elsewhere (Tyndale-Byscoe, 1973; Flint *et al.*, 1981). In this group the blastocyst is prevented from implanting until the lactational stimulus stops when the young leaves the pouch naturally, or because it has died. Under normal circumstances the young are born immature and underdeveloped and require a long lactation in the pouch but after birth the female is inseminated again so that another blastocyst is ready to implant as soon as the lactational stimulus from the pouch ceases. There are exceptions to this stimulus-response situation in species such as the tammar wallaby (*Protemnodon eugenii*) and the quokka (*Setonyx brachyurus*) where, at the end of one breeding season, the ceasing of lactation does not trigger implantation and this waits until the next breeding season (Weir and Rowlands, 1973).

In the eutherian mammals there are at least two explanations for the occurrence of obligate delayed implantation (Weir and Rowlands, 1973). In the stoat (*Mustela erminea*) the female mates in June or July but does not give birth until the following spring. The similar, but smaller, weasel (*Mustela nivalis*) has a pregnancy of about 42 days so there is obviously no need for a long gestation in the stoat and the delay of implantation has been demonstrated to last about 280 days (Deansley, 1943; King, 1977). Young female stoats mate and conceive in the year of their birth, even before weaning in some cases, and it is suggested that implantation is delayed in order that the somatic growth is sufficient to bear the stress of pregnancy. This unusual occurrence of delayed implantation also ensures that most females are fertilized (probably not by their father) before littermates have dispersed. The other explanation is that because birth must occur at the optimal time for the survival of the young, mating must be timed and adapted to provide a pregnancy which terminates at the best time for the young. If, however, the time of mating is restricted to only one time in the year, for example at the post-partum or post-lactational oestrus in seals, then some form of delay is necessary to allow pregnancy to end at the right time, up to a year later.

The other species in which delayed implantation occurs, such as the badger (*Meles meles*) and the roe deer (*Capreolus capreolus*), may be adapted to mating at favourable times of the year as well as giving birth at the optimal time for the survival of the young. Both these species, have a relatively short period of gestation from implantation to birth, January–March or December–February (Asdell, 1964) for the badger and December–May or January–June (Chapman, 1977) for the roe deer, and it seems likely that these species are avoiding an energetically expensive period of activity (reproductive maturation and mating) when food supplies are likely to be poor. Delayed implantation in species such as the mink (*Mustela vison*) is more difficult to account for. Mating occurs early in the spring (four oestrous cycles between February and early April, (H.V. Thompson, 1977) with a minimum period of gestation of 39 days and a maximum of 76 days (average 45–52 days). The period of delay seems to get shorter as spring progresses (Dukelow, 1966) and is related to the daylength, so that increased light after mating advances the time of implantation (Pearson and Enders, 1944). There is also, apparently, no reason why mating should not take place immediately before implantation in the grey seal (*Halichoerus grypus*) as mating can take place at this time (Harrison, 1963) or in the previous autumn, as depicted in Fig. 2.10.

The proximate control of delayed implantation in those species which have a seasonally fixed delay is likely to be photoperiodic (Wimsatt, 1975). Experimental attempts to show that this is the case have succeeded in a number of species, for example, the badger (*Meles meles*), spotted skunk (*Spilogale putorius*), and mink (*Mustel vison*) but not in the roe deer (*Capreolus capreolus*).

Delayed development

It is again in bats that delayed development, either in mid-term or at the end of pregnancy, is shown. The non-hibernating species which have been most studied are the phyllostomid equatorial fruit bats (*Artibeus jamaicensis*) (Fleming, 1971) and leaf-nosed bat (*Macrotus californicus*) (Bradshaw, 1962). A further species reported to show delayed development is the natalid (*Natalis stramineus*) (Wimsatt, 1975). In the leaf-nosed bat fertilization occurs between September and November and embryonic development proceeds slowly through the winter so that the young are born in June. In *Artibeus* there are two births each year in March or April and July or August. A post-partum oestrus occurs after the birth in March or April and a normal pregnancy follows with parturition in July or August. However, after the second birth the female becomes pregnant again and blastocysts implant in August or September but do not begin continuous development until mid-November so that the young are born in March or April. The two periods of birth occur at energetically favourable times of the year (Fig. 2.11) when fruits are most plentiful, rather than, in the case of the second pregnancy, during November–January when fruits are scarce. The adaptive value of this delay in development is obvious, but as Fleming (1971) points out, the proximate factors causing the delay are unknown.

A phenomenon related to delayed development should also be discussed at this point. In heterothermic bats (e.g. many vespertilionids and rhinolophids) it is possible to delay parturition if the temperature is depressed and the food supply is poor (Racey, 1973). Experimental work with pipistrelle bats (*Pipistrellus pipistrellus*) shows that when pregnant bats are deprived of food in a cold environment they become torpid and that

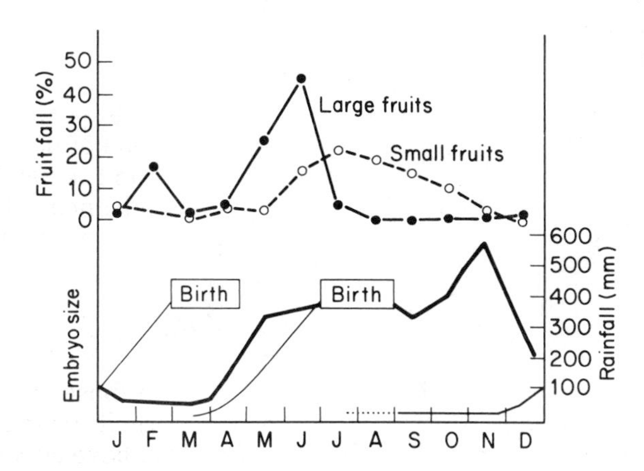

Fig. 2.11 The relationship between rainfall (thick line), availability of large and small fruits and the birth periods of the fruit bat (*Artibeus jamaicensis*) in Panama. Data on rainfall are for Cristobal, on the Atlantic coast of the Panama Canal Zone. Embryos are presumed to be conceived in March and July or August. Redrawn from Fleming (1971). Copyright 1977 by the AAAS.

pregnancy is extended by a period which is similar to that of the induced torpor, indicating that foetal development has been arrested. These experiments were carried out at 5–14 °C but when the bats are placed in temperatures of 10–25 °C, and given food, gestation is normal and when in 30–35 °C the gestation is significantly shortened. This facultative delay in development is obviously adaptive and is simple to explain as the poor environmental conditions are exactly matched to the length of the delay.

Parturition and weaning – strategies for survival in the postnatal environment

There are many variables in the reproductive process of mammals after birth. The dependence of the young on the care and nourishment supplied by the mother means that rates of growth, competition with littermates, time to weaning, interactions with other adults and predators may, among other variables, all affect survival. Litter size, frequency of litters and time to weaning are statistical measures which are more relevant to the discussion in Chapter 4 but the morphological and behavioural variables in the immediate post-weaning period are discussed here.

Precocial and altricial young

The stage of development at birth in mammals is very variable but, generally speaking, it is possible to consider a species as having well-developed or little-developed young. If the young are born so that they are capable of independent locomotion and of keeping warm then they are known a precocial; if the young must be protected in a nest or refuge they are known as altricial (Ewer, 1968). The stage of development at birth is very much related to the mode of life of the species so that burrow-dwelling species of rodents might be expected to have young that are hairless, with non-functional eyes and ears and little power of thermoregulation. In contrast, surface dwelling rodents are likely to be well developed at birth with a covering of hair and functional senses.

Maturity at birth is difficult to define precisely and the nearest measure one can use is that of birth weight in relation to adult body weight. For a particular adult body weight birth weight is likely to be greater the smaller the litter size (Case, 1978; Leitch *et al.*, 1959). This relationship is shown graphically for three groups of rodents in Fig. 2.12. In addition to this relationship McKeown, *et al.* (1976) suggest that as total litter weight is approximately the same irrespective of the number of young, the individual offspring from species with large litters will be relatively smaller than those from small litters; the relatively larger young are also likely to be more mature. Thus odd- and even-toed ungulates have few young, often only one, and they are precocial. Cetaceans have a single young which must swim at birth although it may be dependent upon the female for up to a year for food (milk); the pinnipeds have single young which are weaned in a relatively short time in comparison with their age at maturity. The exceptions are the Chiroptera and the primates which usually have single young but which are

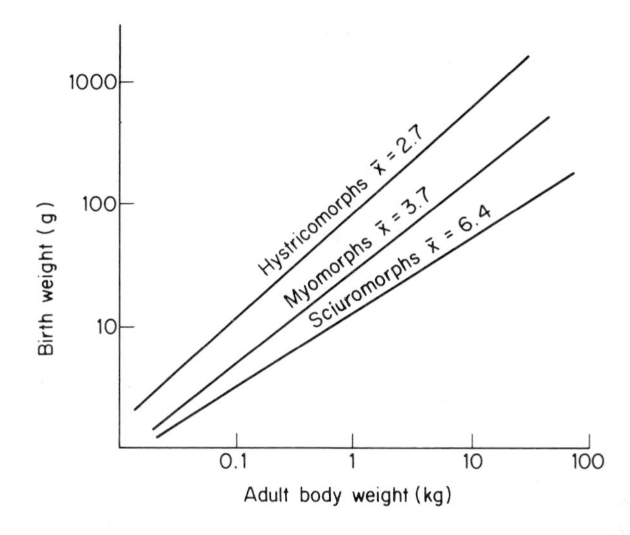

Fig. 2.12 Relationship between birth weight and adult weight in three groups of rodents having marked differences in mean litter size ($\bar{x}$). Redrawn from Case (1978).

usually altricial. McKeown *et al.* (1976) attribute this anomaly to the arboreal or aerial mode of life where it would be difficult for the mother to transport more than a single young. In the fissiped carnivores, insectivores, rodents and lagomorphs the litters are usually greater than one and much use is made of protective burrows and dens for the protection of the altricial young. There are, of course exceptions such as the hares (*Lepus* spp.) which give birth to precocial young and some members of the rodents and carnivores, but the large litter size – altricial young relationship seems to hold true for a large number of species.

Case (1978) considers the advantages of producing altricial young and bases his argument on the assertion of Sacher and Staffeldt (1974) that mammals with altricial young usually have shorter gestation periods than precocial species. Case argues that it is reasonable to assume, given the latter assertion, that the selective pressures operating on the mother will favour altricial offspring whenever the nest site is a safer place for an embryo to complete development than inside a mother's body, or where foraging demands so much skill and experience that the young, no matter how precocial, could not effectively compete with the adults in the population. Species needing foraging skills may be unduly burdened by the longer pregnancy needed to produce precocial young, but if food is easy to find (e.g. grass) they may produce precocial young since a longer pregnancy would not interfere as seriously with the mother's foraging activities. Such precocial young would be able to find and consume such easily obtainable foods soon after birth as the adaptations for consumption depend more on morphology and physiology than on learning to perfect hunting techniques. In the light of these arguments Case concluded that it is not surprising that ungulates, hares, cetaceans, seals and primates are usually more precocial at birth than are mice, squirrels, carnivores and bats.

Behavioural adaptations of parents and young

This subject has been considered at length from both the behavioural (Ewer, 1968) and evolutionary (Trivers, 1974; Alexander, 1974) viewpoint. It is useful, however, to outline some examples of how mothers and their young avoid predation.

The ungulates provide a very good example of two differing strategies in mother–infant relationships and this group's behaviour has been well reviewed by Lent (1974). In ungulates the young are generally precocial and during the immediately post-partum period the mother and young undergo intense interactions involving movements, odour, visual, tactile and vocal stimuli which elicit maternal behaviour in the mother and direct the activities of the young. After this post-partum period there are two distinct forms of maternal-infant behaviour in ungulates (Walther, 1961, 1964, 1965, 1968; Lent, 1974). These are the hider types and the follower types. The hiders are generally species using forested habitats or small species which are able to take advantage of low cover in open habitats; they include most cervids and gazelles and many antelopes. The mother and young remain apart for long periods of time and this phase typically starts with the infant finding its own hiding place in some cover away from the site of parturition. In the goat (*Capra hircus*) and the waterbuck (*Kobus defassa*), however, the mother leads the infant to a new hiding place (Rudge, 1970; Spinage, 1969), and in some other species the mother simply walks away from the calf. Once the infant is separated from the mother further contact is made only at some distance from the hiding place so that the female waits for the emergence of its young rather than walking up to the refuge. Nursing and care sessions may be as infrequent as two or three times per day or as often as ten times per day. In this way the female avoids drawing attention to the infant except at these times and thus helps to avoid the possibility of predation on the infant. Further predator avoidance measures include the wide spacing of infants, the lack of odour of infants and the mother's ingestion of the placenta and the neonate's faeces. At the same time the infant is gradually introduced to a closed social group.

The follower type species include the equids, most large bovines, sheep and related genera, the musk ox (*Ovibus moschatus*) and the caribou (*Rangifer tarandus groenlandicus*). The basic differences between the hider and follower types of behaviour are illustrated in Fig. 2.13. The mother and young of the follower type maintain close contact and the decline in the frequency of contacts after the post-partum period is not so marked as in the hider type. This allows the mother and infant to be more mobile and many follower species are characterized by great seasonal mobility in grassland or tundra habitats. In the follower it is the mother who has to defend the young in the event of attempted predation and the common association with large aggregations of conspecifics will also help in this respect.

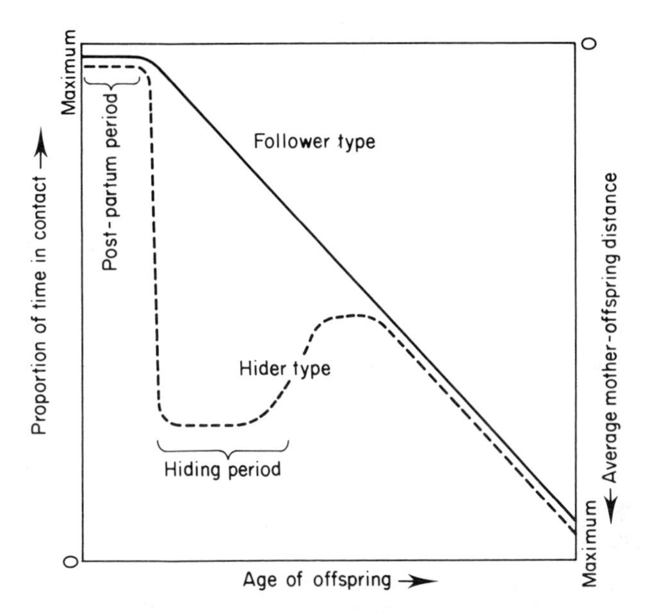

Fig. 2.13 Relationship between contact time and age in the 'hider' and 'follower' types of maternal-infant relationships found in ungulates. Redrawn from Lent (1974).

The different strategies of social and solitary parturition have led to differences in the timing of birth in ungulates and other animals. Gosling (1969) pointed out that in Coke's hartebeeste (*Alcelaphus buselaphus cokei*), which lives in mixed scrub habitat parturition takes place away from other members of the herd and most give birth in shrubland, presumably for camouflage against predators. Parturition takes place at any time of the day and for the first two weeks of life the mother–infant relationship is that of a 'hider'. In the wildebeeste (*Connochaetes taurinus*) however, the females do not isolate themselves prior to parturition and give birth in short grassland amongst other adults. The calves stand four times as quickly as do those of the hartebeeste and they show the 'follower' type of behaviour. Calving occurs in a restricted season before the long rains and the calves experience a heavier rate of predation than those of hartebeeste. However, this is offset by faster development and the effect of the restricted season of births which results in a superabundance of prey which cannot be completely used by the predators (lions and hyaenas). Most parturition in the wildebeeste takes place in the early morning, presumably so that the calves can have all day to learn to stand before the predators are active. The calves are not meticulously cleaned and the expulsion of the placenta is delayed until the calf can move so that predators are not attracted (it is not eaten in this species).

The main strategy for the timing of reproduction in the two species discussed above is similarly related to the rains but after that the details of reproductive behaviour are very different. Because of the different

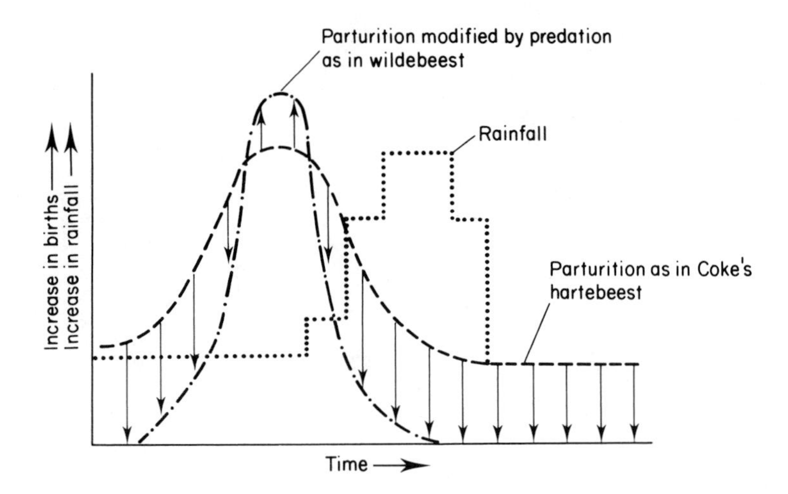

Fig. 2.14 Adaptation in the season of births. The diagram shows how a seasonal distribution of parturition (————) may be adapted to climate (· · · · · ·) but modified by more efficient predation on calves born outside the main parturition season (—·—·—·—). The arrows indicate the effect of predation on different levels of parturition and suggest how one type of distribution might evolve towards the other. Redrawn from Gosling (1979).

habitats and behaviour the effects of predation are quite different and Gosling proposed a model to show how predation could alter the synchrony of the times of parturition (Fig. 2.14). The rains coincide with a period of good food supply and both species are adapted to give birth peaks at this time. However, in the species which suffers much predation it is advantageous to swamp the predator with young at one time rather than spread out the season of births and this is what has apparently happened in the wildebeeste in comparison with the hartebeeste; the latter having a long, less intense period of birth and the former a short intense period of birth. This effect was first noticed by Darling (1938) in connection with sea birds and has been called the 'Fraser Darling Effect'.

These examples show quite vividly that reproductive efforts by no means end with mating and pregnancy and that the female in particular has a great burden energetically and behaviourally before the young are self-sufficient. Feeding strategies, habitats and the incidence of predation all have a part to play in determining the nature of the relationship between mother and young.

3

Effects of the Environment on Reproduction

Mammals show a wide variation in the timing of reproductive activity and in its success during their lives. Major factors affecting this variation at all stages are environmental, such as climate, social influences and nutrition, and many of these effects have been catalogued at length (Sadleir, 1969; Gilmour and Cook, 1981). In this chapter examples of environmental influences on reproduction are drawn from both wild and domestic mammals, and their significance to reproductive periodicity and function will be explored. Such environmental factors have been equated with 'exteroceptive' factors, a term first used to describe external local factors affecting the pituitary gland (Amoroso and Marshall, 1960); however, it is by no means certain that all environmental factors influence reproduction via this organ as behavioural and more direct endocrinological effects are possible.

Reproductive activity encompasses puberty and, if breeding is seasonal, the re-initiation of ovarian cyclicity or testicular function. Once the physiological necessities for reproduction are present then reproductive behaviour, patterns of courtship and mating are necessary for successful fertilization leading to pregnancy and lactation. Any one of these physiological or behavioural processes may be affected by environmental factors. Puberty and the initiation of seasonal breeding may be affected in similar ways by the environment and if the effects are adverse they may be severe so that reproduction is prevented; alternatively the processes involved in reproduction may not be completely inhibited by the changes in the environment but rather may be modified so that they occur at a different time and/or with reduced or increased success.

The evolution of breeding seasons

Given that energetic and nutritional constraints are likely to limit the time when breeding will end in the successful rearing of young, it is not surprising that in seasonal environments reproductive activity should be timed precisely. In many species it is likely that parturition and lactation are coincident with the most favourable time for nutrition of the lactating female and the weaned young so that survivial of both mother and young is as assured as it can be. A cursory study of a range of northern hemisphere mammals (Fig. 3.1) shows that mammals can conceive at any time of the year but that most births occur in the spring or early summer and if births

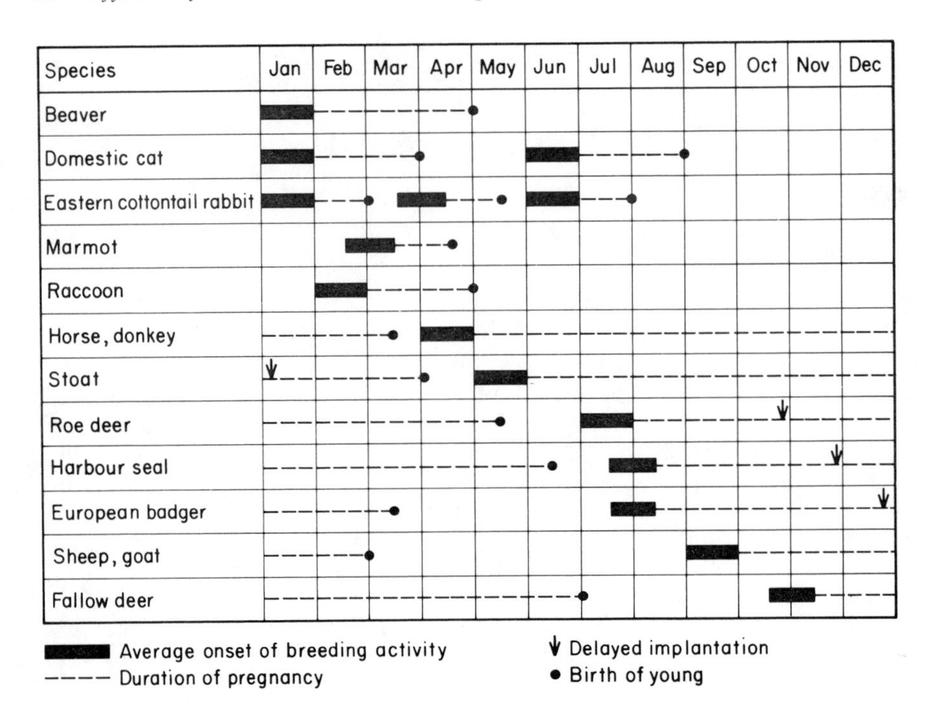

Fig. 3.1 Annual reproductive cycles of some female mammals from the northern hemisphere. Only one complete cycle is shown except for a few species breeding very early; the timetable assumes that pregnancy is established during the first period of behavioural oestrus. Redrawn with modifications from Hoffman (1973).

do occur later in the year they are the result of second pregnancies. However, this pattern of fixed seasonal reproduction is not necessarily common to all mammals, as many species living in unpredictable environments give birth whenever the food supply is available, regardless of season, if there is one. The montane vole (*Microtus montanus*) seems to be such a facultative breeder (Negus *et al.*, 1977); in this species the availability of green plant food has been shown to exert a primary influence on reproduction so that oestrus is induced within 24 hours after a meal containing sprouted wheat or lettuce leaves, and gonadal hypertrophy in males is elicited after plant food supplements are added to the diet. Similar direct effects of food plants are seen in the round-tailed ground squirrel (*Spermophilus tereticaudus*) and probably in many other desert species where the germination of desert annual plants in response to rainfall appears to initiate reproduction (Reynolds and Turkowski, 1972). Species such as the brown rat (*Rattus norvegicus*) living in favourable environments probably never stop breeding after puberty; females in commensal populations show evidence of pregnancy throughout the year (Perry, 1945).

If a species has a short pregnancy it may easily respond to favourable conditions for breeding by maturing gametes, mating, giving birth and lactating within a short period of time. However, many mammals show a relatively long pregnancy which would exceed the length of the favourable period for lactation and weaning. These species may need to mate and become pregnant at the end of a favourable period or during an unfavourable period so that lactation may occur at the best time of the year. Under the latter circumstances species such as the mink, deer and sheep will mate during late summer or autumn during shortening periods of daylight and give birth in the following spring. These 'short day' species contrast with the more conventional 'long day' mammals and other vertebrates which in seasonal environments mate in spring as the days are lengthening. The stimulus inducing maturation and mating may be a change in photoperiod, temperature or food. This is what Baker (1938a) called the 'proximate' factor which triggers reproductive activity. He described the environmental conditions which are optimal during the season of births as the 'ultimate' factors to which the population is adapted. Thus the proximate factors must initiate reproductive activity at a precise time, before the ultimate factors occur, for breeding to be successful; the ultimate factors are likely to include food supply but could also involve climate or predator pressure. Such breeding strategies occur under predictable environmental conditions and are thus only seen in fixed season breeders. Tropical and equatorial species may show a definite rhythm of reproduction (Spinage, 1973; Lofts, 1970), but it is debatable whether they are responding to an external cue or simply following an internal 'endogenous' rhythm of reproduction. At the equator many external cues may be lacking – photoperiodic change, for instance, is minimal; and equatorial bats, for example, appear to breed continuously (Marshall and Corbet, 1959). Such continuous breeding may, however, contain peaks, as in the warthog (*Phacochoerus aethiopicus*), waterbuck (*Kobus ellipsiprymnus*) and gazelles (*Gazella* spp.) (Spinage, 1973), and discontinuous breeding does occur at or near the equator, as shown by various species of flying fox (*Pteropus*) (Amoroso and Marshall, 1960).

The evidence for an endogenous rhythm of reproduction is not overwhelming but has support from studies of sheep and ferrets. In sheep held in constant light oestrus still occurs circannually (Ducker *et al.*, 1973) and in ferrets the eyes are required for the synchronization but not the periodic occurrence of breeding (Herbert *et al.*, 1978). In species such as the golden hamster (*Mesocricetus auratus*) there appears to be no evidence to show that the reproductive rhythm is inherent (Reiter, 1974); under conditions of constant dark (blindness) adult male hamsters undergo regression of the gonads for about 20 weeks followed by regeneration and continued functioning of the reproductive tract up to 63 weeks from the start of the experiment. No further regression of the gonads will take place in the absence of photoperiodic information and so the 'free-running' of any endogenous rhythm is not apparent. The details of photoperiodism in the hamster are discussed later in this chapter.

Although, in many species, the exact timing of reproduction may be under strict environmental control, there is now evidence to show that there are polymorphisms in the reaction of individuals to environmental stimuli. In the bank vole (*Clethrionomys glareolus*) Clark (1972) found that some individuals in winter light conditions do not stop reproduction, suggesting that the population in winter may have apolymorphism for the reproductive reaction to short days; thus winter breeding, which is normally uncommon, may take place in some individuals despite the fact that they are being subjected to normally inhibitory short day lengths. Similar results have been obtained with the white-footed mouse (*Peromyscus leucopus*) (Johnston and Zucker, 1980). Also, in the deer mouse (*Peromyscus maniculatus*) (Fairbairn, 1977a) there is an early peak of female breeding in the spring, before environmental conditions are very favourable and a later peak when conditions are better. The early breeding females risk mortality, apparently because of the poor conditions, but they also gain the opportunity of having a greater number of offspring during that breeding season than females which do not breed early. This dichotomy in breeding strategy is possibly phenotypic, there being some flexibility in the timing of the start of breeding. Fairbairn's and other population studies suggest that the reproductive success of each type is equal and so the divergence in breeding tactics may be maintained for this and possibly other reasons. Early breeders are, however, very variable in their breeding success.

Much of the evidence for the involvment of the environment in the timing of reproduction comes from laboratory studies on a few species of mammals; some of these studies are considered below.

Photoperiod and reproduction

Once mating has taken place changes in day length have only a minor modifying role to play in reproduction. In a review of the environmental factors affecting the length of gestation Racey (1981) concluded that changes in the proportion and periodicity of light extend the length of gestation in some laboratory rodents; particular species tend to give birth in light or in darkness, and this may have a slight effect on the length of gestation; he also concluded that seasonal changes in photoperiod may account for variations in gestation length in domestic species and in other species photoperiod may have an effect on the timing of delayed implantation. In contrast to these minor environmental influences on gestation, photoperiod can, in some circumstances, initiate or prevent reproduction according to the species' reaction to long or short days.

Evidence for the photoperiodic control of reproduction

The change in day-length with season in non-equatorial areas is very marked (Fig. 3.2). Thus it is not surprising that many mammals appear to use this precise change in day-length to trigger reproductive activity; many species start breeding as soon as the day-length reaches a critical time. Brown hares (*Lepus europaeus*) studied in six countries in the northern and southern hemispheres started breeding soon after the shortest day of the year (Flux, 1965) and the primitive Soay sheep from the island of St Kilda off north-west Scotland regularly rut (mate) in October–November and give

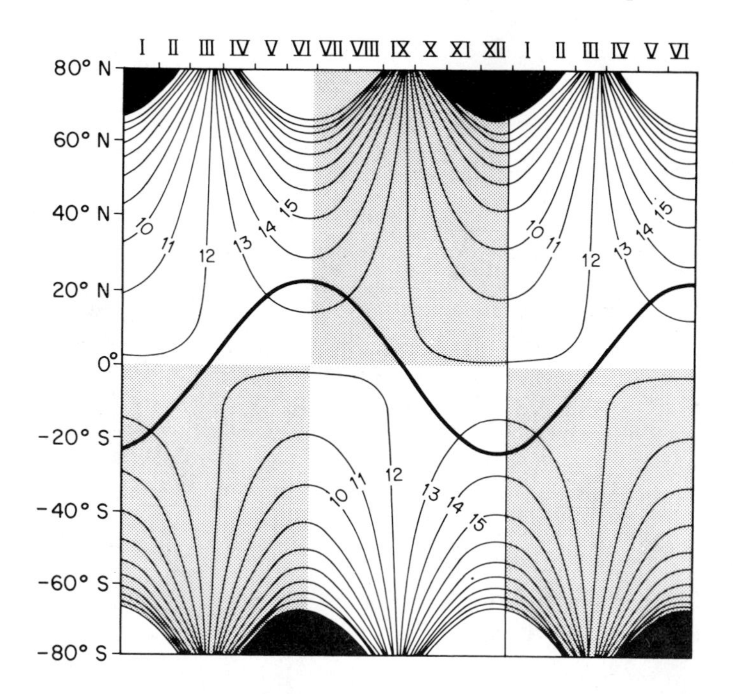

Fig. 3.2 Seasonal changes in the position of the overhead sun (thick black line) and in the length of day (numbered curves). The numbers indicate the hours of daylight. The stippled area represents decreasing day-length and blacked-in areas show when the sun is completely below the horizon for 24 hours. Redrawn from Baker (1938b). By permission of the Zoological Society of London.

birth at a marked peak in April (Grubb and Jewell, 1973; P.A. Jewell, personal communication): median dates of birth from one study area were April 18 (1965), April 14 (1966) and April 12 (1967) followed by April 19/20 (1978), 22/23 (1979), 23 (1980), 12 (1981), 25 (1982), 22 (1983), and 23 (1984). Synchrony of reproduction repeated every year does not, however, prove that changes in photoperiod are initiating reproduction; further evidence is needed.

Many seasonally breeding species shift their breeding season by six months when translocated across the equator, providing circumstantial evidence that photoperiod is involved in the timing of breeding. Thus ferrets (*Mustela furo*) which were taken from Britain to South Africa adjusted their breeding season to fit in with southern hemisphere light regimes (Bedford and Marshall, 1942); similarly, fallow deer (*Dama dama*) from England, moose (*Alces alces*) and wapiti (*Cervus canadensis*) from Canada, white-tailed deer (*Odocoileus virginianus*) from USA and chamois (*Rupicapra rupicapra*) from Europe all reversed their breeding season from September–November (according to species) to March–May when transported to New Zealand (Marshall, 1937, 1942). Also, the breeding season of many species varies with latitude. Bank voles (*Clethrionomys glareolus*) start breeding in April in England and Wales and the majority of females are pregnant in May, whereas these events occur in May and June in the more northerly Scottish populations (Delany and Bishop, 1960; Brambell and Rowlands, 1936). However, an Island population of bank voles off west Wales shows a similar adaptation at the same latitude as the mainland population (Jewell, 1966).

More precise and convincing evidence concerning the effect of photoperiod on reproduction comes from experimental studies where the lighting regime is manipulated to see its effect on reproduction. Many such experimental studies are reviewed by Sadleir (1969). A summary of experimental studies of ferrets is given in Fig. 3.3, showing how popular this type of experiment has been and also how confusing the results appear! One interpretation of all the various results is that if ferrets are under the influence of light at around 12 hours after a dawn stimulus then they show reproductive activity, irrespective of whether there is darkness between the 'dawn' and the later light stimulus. This mechanism of light stimulation is discussed on page 67.

Experimental studies of voles were pioneered by Baker and Ranson (1932) who worked with the field vole (*Microtus agrestis*). They compared vole reproduction under laboratory light regimes of 15 hours of light and 9 hours of dark (15L:9D) and 9L:15D. Reproduction was normal in the 'long

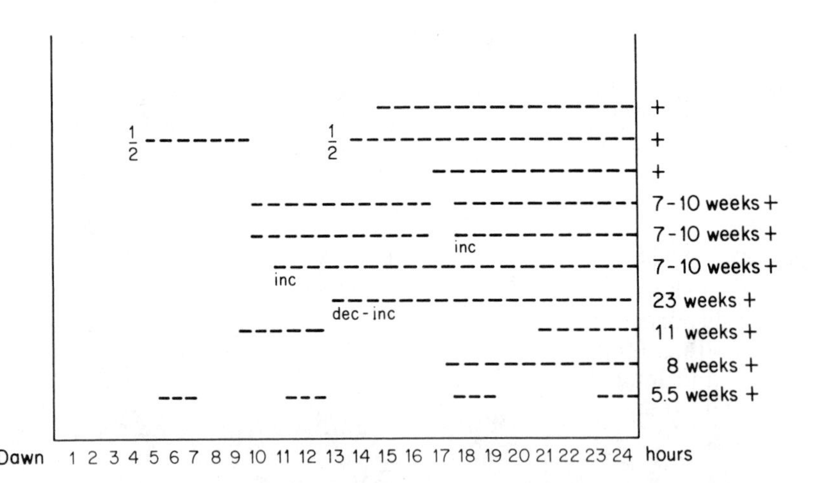

Fig. 3.3 Experiments on the effects of various day-lengths on breeding (oestrus) in the female ferret. It is assumed that a photo-inducible phase probably lasts from 12–20 hours after dawn (the start of the primary light period). According to the external coincidence model (see text and Fig. 3.8) it does not matter how long the first (dawn) period of light lasts as long as it entrains the photo-inducible phase for 12–20 hours later, the results of many of the experiments on photoperiod and ferret breeding may be explained by this model. Details of the experiments and references are given in Sadleir (1969). Note that – denotes a period of darkness, ($\frac{1}{2}$ = half an hour of darkness), + = breeding occurred (after time, if stated), inc = increasing light, dec = decreasing light.

day' group but it was almost completely curtailed in the 'short day' group. Comparison of six different photoperiods and their effect on reproduction (Breed and Clarke, 1970) showed that in voles there appears to be a threshold photoperiod of between 12 and 14 hours of light above which sexual development of 18 day old females is more rapid than in photoperiods below this level; perforation of the vagina took 42 days in the 12L:12D group but only 17 days in the 14L:10D group.

In the golden hamster (*Mesocricetus auratus*) extensive studies of the effects of photoperiod on reproduction have answered many questions concerning mammalian photoperiodism (Hoffman, 1973; Elliott, 1976; Reiter, 1974, Turek and Campbell, 1979). Hamsters are nocturnal and they hibernate throughout the winter months. Short days (less than 12.5 hours of light per 24 hours) stimulate gonadal regression as the animal enters hibernation. However, in the hibernatory state there is a spontaneous recrudescence of the gonads so that they are fully functional on arousal in spring, allowing successful breeding immediately. Puberty is independent of photoperiod but once puberty is attained the animals become photoperiodic and short days (as above) will induce gonadal regression. Normally long days (12.5 hours of light or more per 24 hours) are needed to maintain reproductive activity. Long days will stimulate reproduction before natural recrudescence, so it is inferred that the individuals remain photoperiodic during hibernation, but the natural short-day light regime (and perhaps the individual's distance from light during hibernation) prevents development before the endogenous timing mechanism initiates recrudescence of the gonads. However, once the testes have developed spontaneously after short-day induced regression they will not respond to short days again until the animal is subjected to a period of long days. This photorefractory state is only terminated by 10–22 weeks exposure to long photoperiods. (Note that this is the opposite situation to that observed in the refractory state of birds (Follett, 1973, 1981) where, after stimulation by long days in spring, the gonads will spontaneously regress and only respond to long days again after a period of short days has occurred, normally over the next winter.)

Short-day breeders such as the sheep have also been the subject of extensive study (e.g. Lincoln and Short, 1980) and the alternation of short days (8L:16D) and long days (16L:8D) at 16 week intervals (making 32 week 'years') will induce testicular growth to occur every 32 weeks instead of every 52 weeks as under normal light conditions (Fig. 3.4).

Thus photoperiodic changes seem to be potent stimuli influencing mammalian reproduction; how photoperiod is translated into physiological change is discussed in the next sections.

Day-length measurement in mammals

The understanding of how mammals measure photoperiodic changes in the environment owes much to studies of birds (Follett, 1973, 1978) and the lead taken from Bünning's (1936) observation that photoperiodic changes can synchronize the endogenous rhythm of plant leaf movements and the flowering response to light. This led Bünning to put forward the hypothesis that the physiological basis of photoperiodism in plants may involve the

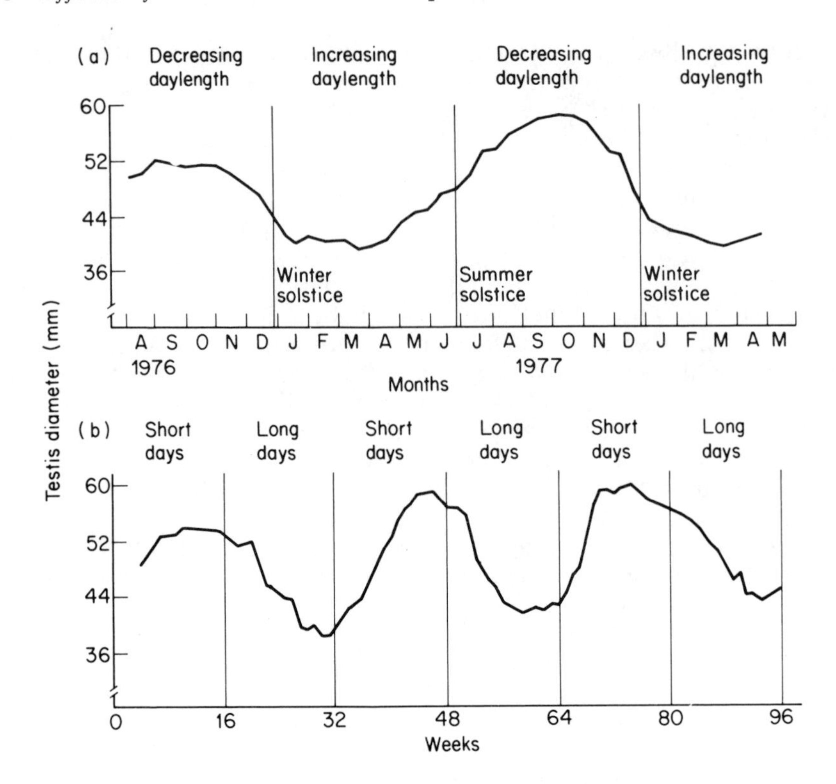

Fig. 3.4 Graphical representation of the changes in the mean diameter of the testes of 14 Soay rams **(a)** living outdoors under natural lighting near Edinburgh and **(b)** living under experimental lighting conditions of alternating 16-week periods of short days (8L:16D) and long days (16L:8D). All rams were 1 year old at the start of the experiment and were from the same flock. Redrawn from Lincoln and Short (1980).

organism's circadian rhythms (see Bünning, 1967). Although Sadleir's (1969) review puts this idea forward as one of a number of possible physiological mechanisms of mammalian time measurement it was not until Elliott *et al.*, (1972) started doing experiments on the subject with golden hamsters (*Mesocricetus auratus*) that Bünning's ideas were shown to be relevant to mammals. Earlier theories reviewed by Elliott (1976) and Turek and Campbell (1979) had suggested (a) that the total duration of activity, not light, was responsible for gonadal development in response to light stimuli, (b) that the relative ratio of light to dark as compared to the previous ratio is the determining factor in initiating reproduction, or (c) that the total length of the light or dark period is measured by a type of hourglass mechanism which has a critical length of light or dark or critical ratio of light to dark. None of these theories could adequately explain light measurement and it was only when it was realized that the important factor was not *how much* light was present but *when* it was present (in conformity with Bünning's hypothesis) that mammalian light measurement was better understood.

In order to show that a circadian clock is involved in measuring day-length it is necessary to show that it is not important how much light there is in the day but rather when the light stimulus occurs during the day. A common experiment used to demonstrate this phenomenon is the 'resonance' experiment where a relatively short light period is added to different lengths of darkness. The dark periods are kept the same for each of the experiments and are usually fractions or multiples of 24 hours. Thus it is assumed that a period of photosensitivity maintains its 24 hour rhythm and that the period of light will fall within this sensitive phase or outside it according to the length of the dark period. Obviously with experimental

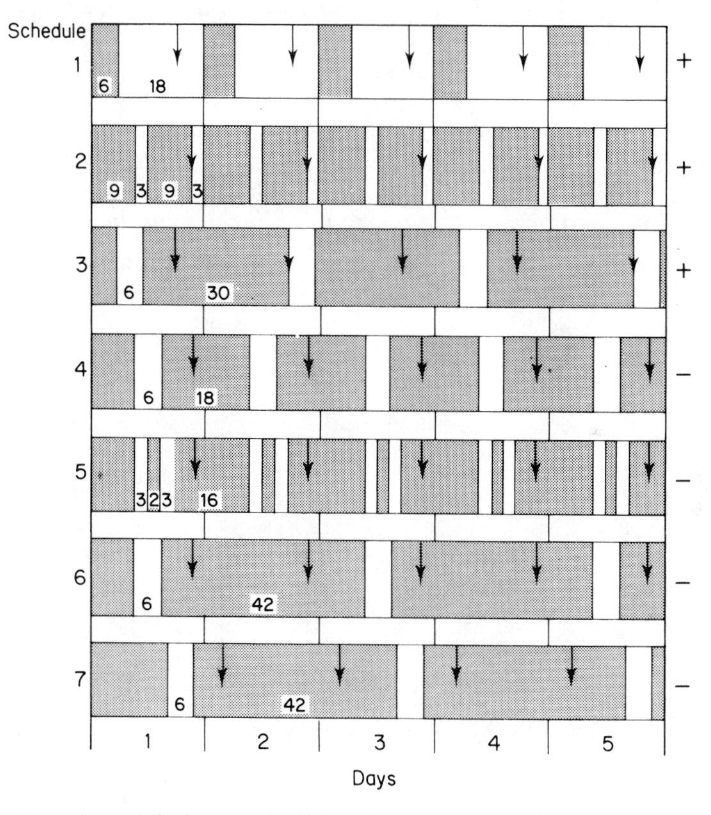

Fig. 3.5 Experimental ahemeral light regimes and reproduction in the field vole (*Microtus agrestis*). The diagram illustrates seven different experimental lighting schedules over 5 days. Light periods are indicated by unshaded areas and dark periods by shaded areas. The figures within the diagram indicate the length of the period in hours. The arrows indicate the approximate time of onset of a postulated photosensitive period (see details of external coincidence model in text and Fig. 3.8) assuming that its position is determined by, and occurs 12 hours after, the time of the experimental dawn. Experiments 1, 2 and 3 permit greater testicular activity than Schedules 4, 5, 6 and 7. Redrawn from Grocock and Clarke (1974).

regimes that have a long period of darkness the period of light might fall in the sensitive phase of the circadian rhythm only once every few days. An experiment such as this was carried out by Grocock and Clarke (1974) on *Microtus agrestis*, the field vole, and the relative development under the different light regimes can be interpreted as showing that as long as light is present at about 12 hours after a 'dawn' then reproductive activity is stimulated, even if the light occurs only once every 36 hours for 6 hours (Fig. 3.5). An early experiment of this type on mammals was on the golden hamster (Elliott *et al.*, 1972) where 6 hours of light in cycle lengths of 24, 36, 48 and 60 hours respectively maintained testicular development only in the 36 and 60 hour schedules (Fig. 3.6). In the 24 and 48 hour schedules light is present in the non-sensitive phase of the circadian rhythm of sensitivity whereas in the 36 and 60 hour schedules light is present in the sensitive phase of the rhythm and so maintains testicular weight. Further experiments by Elliott (1976) on hamsters showed that the sensitive period lasted from just before the start of activity for about 9 hours.

Another way of demonstrating that there is likely to be a circadian rhythm in day-length measurement is to interrupt a long dark period with a short period of light at varying times after the 'dawn' change in light and so compare reproductive activity under varying 'skeleton' long days with a

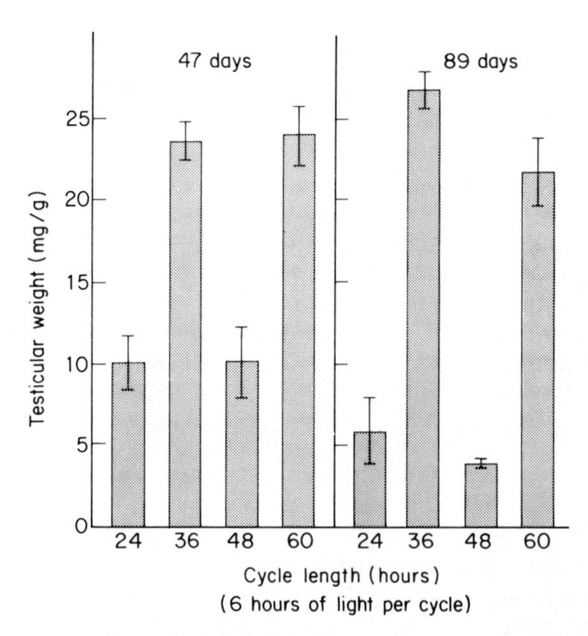

Fig. 3.6 The response of hamster testes to experimental light cycles after 47 and 89 days of treatment. The height of each histogram indicates the mean for the group and vertical bars show the standard error of the mean. All the differences between groups with small testes and those with large testes are significant ($P < 0.05$, Students's *t*-test). Redrawn from Elliott *et al.* (1972). Copyright 1982 by the AAAS.

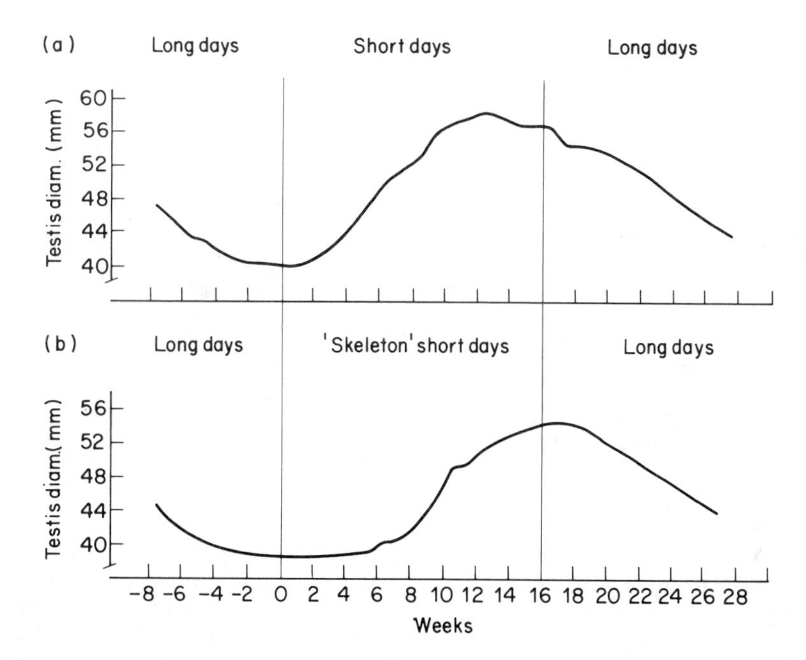

Fig. 3.7 Weekly measurements of the mean diameter of the testes in 8 Soay rams exposed to experimental photoperiods. **(a)** Short-day experiment: 16 weeks of long days (16L:8D) followed by 16 weeks of short days (8L:16D) and then a return to long days. **(b)** 'Skeleton' short-day experiment: 16 weeks of long days (16L:8D) followed by 16 weeks of 'skeleton' short days (11L:1D:5L:7D) and then a return to long days. Redrawn from Lincoln (1978).

total duration of light similar to short days. Similarly, in testing short-day species a long light period can be interrupted with a short period of darkness at varying times after the dawn change in light and so produce 'skeleton' short days with a total duration of light similar to long days. This latter type of experiment was conducted with 'short-day' Soay rams by Lincoln (1978); the skeleton short days (Fig. 3.7a, b) were composed of 11L:1D:5L:7D following long days of 16L:8D and the regime induced testicular recrudescence (which can also occur under prolonged long days (Lincoln and Davidson, 1977)). Thus the total amount of light or dark seems not to be important but, again the timing of the light or dark periods in a rhythm of sensitivity is critical. It seems likely that circadian rhythms are involved in time measurement in many mammalian species (Zucker *et al.*, 1980), however, many details are still uncertain.

There are two interpretations of the various light schedule experiments. One is the 'external coincidence model' (Fig. 3.8a) which predicts that photoperiodic induction of reproduction only occurs when the light stimulus is coincident (or not coincident) with a particular phase of a circadian rhythm of light sensitivity. Here light acts to entrain the rhythm of sensitivity and also to induce a response. The other interpretation is the 'internal

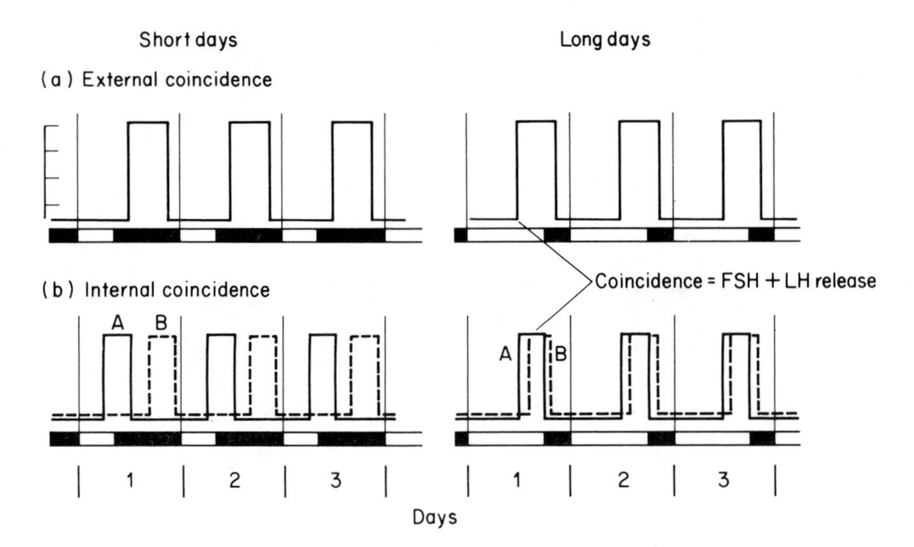

Fig. 3.8 Two possible models of how circadian rhythms might be involved in photo-periodic time measurement in species sensitive to long days. **(a)** The 'External Coincidence' model envisages an organism possessing a circadian rhythm of photo-sensitivity with a photo-inducible phase. Coincidence between this rhythm and light – as occurs under long but not short days – allows reproductive activity. **(b)** One type of 'Internal Coincidence' model. Photoperiodic induction of reproduction occurs with the coincidence of two separate circadian oscillators under long days. The changes in the relative positions of the rhythms (A and B) arise because of the different entrainment which occurs under long days from that under short days. Redrawn from Follett (1978), courtesy of B.K. Follett.

coincidence model' (Fig. 3.8b) which predicts that photoperiodic induction of reproduction only occurs when two or more separate internal rhythms are in a particular phase relationship with each other. Here the light only entrains the rhythms. Which of the two is correct is open to question (Follett, 1978). A further complication is the possibility that the photoinducible phase may alter its position in the 24 hour cycle according to season. This has been suggested for the field vole (Grocock, 1981) where skeleton photoperiods indicate that there is phase shifting of the photoinducible phase between 12 and 16 hours after dawn according to the length and type of dark period. This mechanism would ensure that individuals were not stimulated towards the end of the breeding season although the day-length was the same as that which stimulated the individual in spring.

Despite the lack of complete understanding of day-length measurement the study of mammalian photoperiodism is catching up with studies of birds, and research is starting to elucidate the intriguing question of how short-day species respond to a photoperiod which is the reverse of the normally stimulatory day-length of birds and many mammals. Recent studies (Grocock, 1979, 1980) indicate that the spontaneous growth of testes under normally inhibitory photoperiods is not as uncommon as may have been thought. In the field vole there is apparently a period of photorefractoriness to short days in the winter so that development of gonads occurs spontaneously in individuals of all ages and states of inhibition, and only a short period of stimulatory day-lengths after the equinox in spring is needed to bring about a rapid increase in testicular function.

Physiological effects of changes in day-length

In order that changes in photoperiod should alter reproductive physiology they must be perceived by the nervous system, transmitted to an effector and the effector must respond to the stimulus and affect reproductive physiology and behaviour, via the neuroendocrine and endocrine systems.

Light reception in mammals is generally via the retina of the eye although in neonatal rats non-retineal pathways are proposed (Wetterberg *et al.*, 1970); the alternatives to the retina should not be ruled out in other species. In birds non-retineal receptors in the brain are the rule (Turek and Campbell, 1979). There is a definite neural pathway (retinohypothalamic tract) from the retina to the hypothalamic suprachiasmatic nucleus in the hamster and it has been suggested that the nucleus suprachiasmaticus is the biological clock involved in the hamster's photoperiodic time measurement system (Stetson and Watson-Whitmyre, 1976). Whether the function of the suprachiasmatic nuclei is the same in other mammals is uncertain but similar neural pathways have been observed in species such as the ferret (Thorpe and Herbert, 1976). It has also been observed that there are neural connections between the suprachiasmatic nuclei and the pineal gland in the rat (Axelrod, 1974) and as the pineal produces anti-gonadal substances it may be that this is the final link in the retineal-hypothalamic pathway which affects gonadotrophin releasing hormone (GnRH) and hence reproductive physiology.

Role of the pineal gland

The pineal gland produces a number of anti-gonadal substances including melatonin, an indole on which much work has been done on the assumption that it is the most important of the pineal products affecting reproductive physiology. This assumption is by no means certain but much convincing evidence has been produced to show that melatonin can have an effect on reproductive biology.

Gonadal function in long-day species subjected to short-day light regimes has been maintained by removal of the pineal (Fig. 3.9) and by removal of the nervous innervation to the pineal both physically and chemically (Follett, 1978; Turek and Campbell, 1979). This is not a simple system of pineal inhibition of reproduction, however, as pinealectomy has also prevented gonad growth in the Djungarian hamster (*Phodopus sungorus*) and ferret, and melatonin injections induce testicular growth in golden hamsters in short days (Turek and Losee, 1978). To make the story even more complicated it has also been shown that in the golden hamster injections of melatonin in the late afternoon of an 14L:10D light cycle induced gonadal atrophy whereas injections in the morning have no detectable effect on reproduction (testicular size) (Tamarkin *et al.*, 1976).

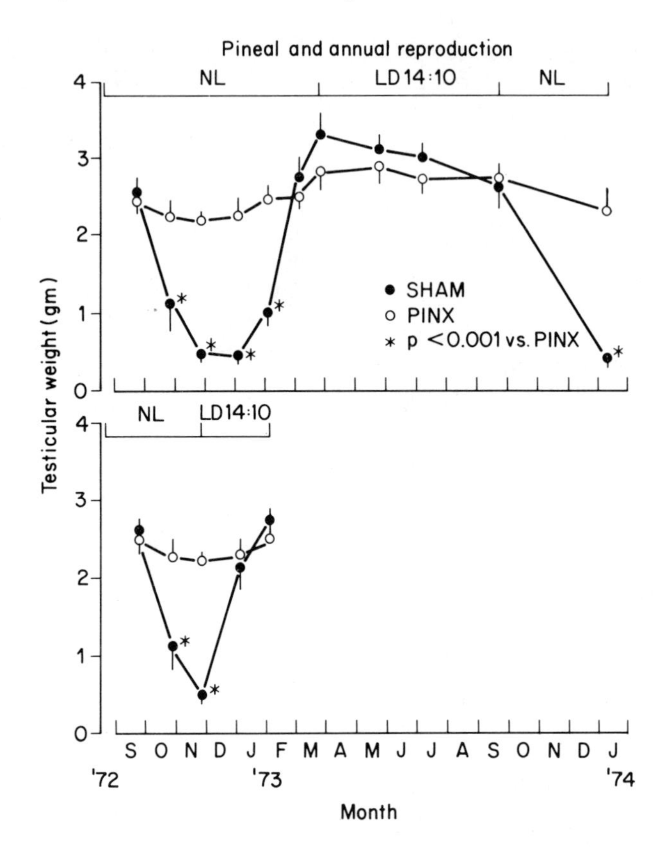

Fig. 3.9 Mean testicular weight of sham operated (SHAM) and pinealectomized (PINX) male golden hamsters kept in groups of 8–12 under natural lighting (NL) conditions or under cycles of 14 hours of light alternating with 10 hours darkness (14L:10D). Vertical lines from points signify standard errors. Redrawn from Reiter (1975).

Melatonin action and interaction in affecting reproductive physiology may be being confused by the other pineal products. However, many species react to melatonin showing an anti-gonadal effect, for example ferrets, weasels, mice, monkeys, rats, gerbils, rabbits and hamsters (see Turek and Campbell, 1979 for references) but the severity of the gonadal regression induced by melatonin varies between species. In the hamster and grasshopper mouse (*Onychomys torridus*), species which are relatively photoperiodic, marked gonadal regression occurred after implantation of melatonin capsules under long day conditions, whereas the rat and the house mouse, relatively non-photoperiodic species, showed no demonstrable effects of the implants (Turek *et al.*, 1976).

In all species studied melatonin production is greater during the night than during the day and its production is controlled by the enzyme N-acetyltransferase (NAT) in the pineal. This enzyme is in turn under the control of the sympathetic nervous system innervating the pineal parenchymal cells (Axelrod, 1974) and both darkness and the proper setting of an internal clock are necessary for the night-time increase in N-acetyltransferase activity. NAT activity is prevented by light during the night but not by darkness during the day in the rat (Axelrod, 1974; Klein and Weller, 1972) and so the scheme of the previously outlined circadian rhythm of photosensitivity seems to fit the scheme of sensitivity of the NAT enzyme to light. Also, Hoffman *et al.* (1981), working on the Djungarian hamster have shown that 1 minute of light in the middle of the dark period in a 8L:6D photoperiod is similar to 16L:8D in stimulating NAT activity only during the 8 hour 'dark phase' and the light regime is thus interpreted in both cases as a 'long day' leading to rapid gonadal recrudescence.

In a short-day species, the Soay sheep, rams subjected to removal of the

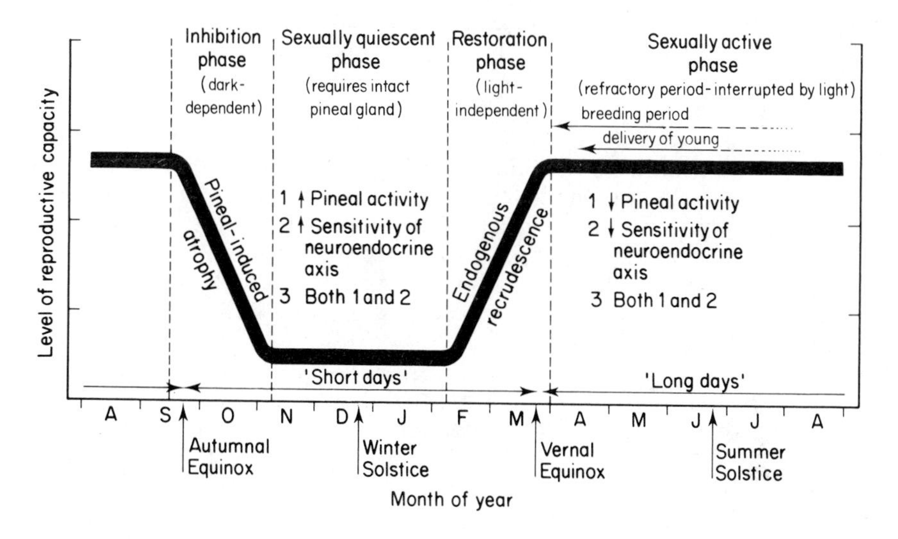

Fig. 3.10 Theoretical relationships between changing photoperiod, the pineal gland and circannual reproductive cycles in a long-day breeding rodent – the golden hamster (*Mesocricetus auratus*) – as proposed by Reiter (1978). The annual cycle of reproduction is divided into 4 phases: inhibition; sexually quiescent; restoration; and sexually active. The inhibition phase is directly dependent on shortening days suppressing reproductive physiology via the antigonadotrophic activity of the pineal gland. In the sexually quiescent phase, the short photoperiods and the pineal gland continue to inhibit reproduction. As spring approaches, reproductive activity is restored (restoration phase): this happens regardless of the day-length and is spontaneous or endogenous. During the sexually active phase, the pituitary-gonadal axis is refractory to the antigonadotrophic hormones of the pineal gland. As autumn approaches, the refractoriness is interrupted and shortening days again induce gonadal atrophy. This annual reproductive cycle is inextricably linked to day-length and the pineal gland. Animals lacking their pineal gland (due to surgical pinealectomy) never experience gonadal involution during the autumn and winter months, even though they are exposed to short days (see Fig. 3.9). Redrawn from Reiter (1978). © S. Karger AG, Basel.

suprachiasmatic nuclei, and thus having non-functional pineal glands (Lincoln, 1978), showed no testicular response to long or short days. In the reports of work carried out on the plasma melatonin levels of control rams (Lincoln and Short, 1980) peak values were recorded in the dark periods but larger and longer nocturnal increases in plasma melatonin levels occurred in the short dark period of 'long day' photoperiods compared with 'short day' photoperiods. This change from short day to long day timing and concentration of melatonin production may explain seasonal breeding in the ram. Recent work suggests that rams become photorefractory at the end of the breeding season, and so reproduction ceases rather than being inhibited by short days (Almeida and Lincoln, 1984).

It is possible that melatonin is involved in the photoperiodic control of reproduction in many species but detailed proof is lacking. In the hamster the theoretical relationships between day-length, reproductive capacity and pineal function have been summarized by Reiter (1974, 1975, 1980) and are depicted in Fig. 3.10; it is likely that many other long-day species show similar interrelationships. The inhibition phase is dependent upon short days as they are necessary for the stimulation of pineal anti-gonadal activity. In the sexually quiescent phase short days and pineal substances still inhibit reproduction and in the restoration phase reproductive activity is initiated regardless of the photoperiod as it is the result of a spontaneous response occurring after a certain period of time of exposure to short days; it is not related to the increasing length of the photoperiod. In the sexually active phase the pineal anti-gonadal hormones are showing reduced activity under long days and the pituitary-gonadal axis may also become refractory to the pineal hormones.

Photoperiodic induction of reproduction

The link between the pineal hormones and the hypothalamo-gonadal axis has yet to be fully elucidated. In the hamster it has been found that short days for 13 weeks increased hypothalamic LHRF (GnRH) (see p. 12) content and reduced serum LHRF and pituitary LH levels compared to individuals on long days (Pickard, 1977); thus it seems that short days inhibit LHRH release. Also, the response whereby the pituitary releases LH and FSH under GnRH stimulation was the same on long or short days and thus the sensitivity of the anterior pituitary to GnRH remained the same regardless of photoperiod (Turek *et al.*, 1977). How the pineal might affect the hypothalamus is not known although pineal factors may affect sensitivity to the feedback of gonadal steroids, particularly oestrogens, regulating the secretion of gonadotrophins (Hoffman, 1973; Karsch and Foster, 1981; Karsch *et al.*, 1980). The subject is still in its infancy and many of the difficulties may be explained by breed or species differences; there is still much to be learned about the control of reproduction in both long- and short-day species of mammals.

Temperature effects on the breeding season and reproductive physiology

The ways in which temperature changes may affect reproduction have been studied mainly in domestic and laboratory species under extreme conditions. Although this environmental variable was widely held to be the major regulator of breeding before the significance of photoperiod was realized (Perry, 1971), little systematic work has been carried out on non-domestic species and often climatic variables are difficult to separate from each other and from their effects on nutrition. As Sadleir (1969) pointed out, it is somewhat academic to separate climate from nutrition and the present physiological condition of an individual will be important in determining the magnitude of effect climate will have on that individual. If temperature acts to advance or retard the start of breeding it may do this by altering a non-reproductive process such as growth and then act through this process (Sadleir, 1972) or it may affect the rate of change of the maturation process and so alter the timing of adult sexual and reproductive function (Piacsek and Nazian, 1981).

Experimental studies of the timing of puberty (vaginal opening) in female house mice have shown that puberty was reached at 33 days of age at −3°C and at 26 days at 21°C (Barnett and Coleman, 1959; Barnett, 1962). Similarly, in female laboratory rats maturation usually occurred more slowly in individuals exposed to 10°C than in those at 22°C. At the other end of the temperature scale there is some difficulty in finding results to complement those at low temperatures. Groups of mice kept at 18°C, 25°C and 32°C showed no significant difference in the age of puberty (Knudsen, 1962) and laboratory rats kept at 30°C showed no difference in the time of vaginal opening compared with controls held at 22°C (Piacsek and Nazian, 1981). Under more natural conditions rats (*Rattus norvegicus*) in an Alaskan rubbish dump showed no difference in the age or size at puberty compared with populations from a more southerly situation (Schiller, 1956).

In North American *Ovis* (wild sheep) species the lambing season varies in date and duration according to the severity of the climate, and presumably some of this variation is caused by differences in temperature. In the northerly latitudes and at high altitudes the lambing seasons are short and come later in the spring than in the warm uplands (Geist, 1971): in the hot deserts of the United States bighorn sheep may lamb in any month of the year. However, the determining factors for sheep breeding seasons may be more complicated than these statements suggest. On examining a large number of studies of North American *Ovis*, Bunnell (1982) concluded that there is little evidence to suggest that the survival of lambs of the desert forms (30–40°N) is influenced by thermal stress, however, in alpine forms (at 40–65°N) factors affecting the mating period and the survival of lambs seem to be involved in the evolution of the breeding season. Most alpine forms lamb over about 2 oestrous cycles (29.8 ± 2.4 days) and their gregarious nature reduces the likelihood of late conceptions. The timing of breeding for most North America *Ovis* appears to have evolved so that lambing occurs outside the period of extreme cold when feeding conditions are poor; in the alpine forms late lambing would necessitate lactation during a period when the vegetation contains a higher fibre content and is of a lower digestibility as well as reducing the period for lamb growth before the winter; thus late breeding would be selected against.

More direct influences of temperature on the breeding season are apparent from studies of species such as the ground squirrel (*Citellus tridecemlineatus*); in this species decreasing temperature is correlated with the initiation of spermatogenesis and males are fully fecund immediately after hibernation, (Wells and Zalesky, 1940); experimental high temperatures were detrimental to reproductive activity. In the coypu or nutria (*Myocastor coypus*), a species which was introduced to the UK from South America in the 1930s and which spread through the wetlands of East Anglia (now much reduced by control operations), reproduction normally continued throughout the year. In this feral population cold weather caused a depletion of fat reserves and as a result of critically reduced nutrient supplies abortion and reduced frequency of littering occurred in the spring months (Gosling, 1981). This is a further example where environmental factors such as weather are difficult to separate from their interaction with food supply and its availability; more severe weather in winter led to extensive adult mortality in the coypu (Newson, 1966). Mating behaviour may also be affected by low temperature. In a study of the cottontail rabbit (*Sylvilagus floridanus*) in 1958 and 1959 the first conception noted had a median date of 26 and 27 February but in 1960 it was delayed until 25 March. This delay coincided with a period of cold weather from 17 February to 27 March 1960 (Wight and Conaway, 1961). Observations of the behaviour of the rabbits showed a 7 day cycle of social interactions related to reproductive behaviour, and in cold conditions social interacts were inhibited because individuals either sheltered or fed intensively (Marsden and Conaway, 1963).

High temperatures may affect the ability of a mammal to feed properly or they may act directly to impair reproductive processes. In the red kangaroo (*Megaleia rufa*), which usually breeds continuously, high temperatures force individuals to seek the shade and thus allow only a restricted time for feeding. The high temperatures coincide with a drought and poor feeding conditions and the result of this temperature and nutritional stress is to impair spermatogenesis and probably to reduce libido (Newsome, 1973); as a result pregnancy rates decline to nil in drought conditions. Reduced fertility in hot weather seems to be common in mammals. Boars show a reduced fertility during and immediately after hot summer weather (e.g. Thibault *et al.*, 1966), but European boars are able to adapt to a tropical climate and maintain normal semen production (Egbunike and Elemo, 1978). In an experimental study the severity of changes in boar semen quality were generally directly related to the duration of exposure to heat (up to about 33–37°C for 6 hours per day over 4.5–7 days); however, in some individuals the incidence of abnormal sperm heads and tails was related to rectal temperature regardless of exposure time; thus heat stress produced an acute rise in temperature which had a greater effect on semen quality than did the duration of exposure (Cameron and Blackshaw, 1980).

The length of gestation may also be altered by changes in environmental temperature; bats seem to be particularly susceptible to this influence, and food deprivation is also involved. In the pipistrelle bat (*Pipistrellus pipistrellus*) foetal development is significantly retarded (by 5 days) if pregnant females are maintained at 5°C with ad lib food for 14–17 days (Racey, 1973); bats maintained at 30–35°C showed advanced dates of birth indicating acceleration of foetal development. Gestation is also delayed during torpor, or adaptive hypothermia, caused by the withdrawal of food to the bats when kept at 5, 10 and 11–14°C. Racey (1981) suggests that torpor may enable other heterothermic species to survive a shortage of food during pregnancy and this could possibly occur in the pocket mouse (*Perognathus*) and deer mouse (*Peromyscus*) species as well as in the hedgehog (*Erinaceus europaeus*) in response to food shortage or low temperature.

Nutritional factors in reproduction

It is clear that nutritional factors are bound up with other environmental influences such as temperature and other climatic variables in the way that they influence reproductive physiology. Also, if food hunting or gathering is prevented or curtailed then nutrition may be reduced despite the abundance of food. Nutritional factors are also intricately linked with trace elements and vitamins as well as levels of reproduction inhibiting or stimulating compounds; these latter may be the so-called 'plant oestrogens' and other hormone-like factors.

From studies of domestic mammals (e.g. Rattray, 1977) it seems that variation in nutrition can have profound effects on almost all stages of reproduction from puberty, sexual development and reproductive efficiency to prenatal and postnatal foetal survival. In sheep and cattle as well as in laboratory rodents size appears to be more important than age in determining the time of onset of puberty; the growth rate will thus affect the age at which puberty is reached. A study of rabbits (Myers and Poole, 1962) supports this close relationship between body weight and puberty; females reached puberty at 1200 g at 7 months of age but under low nutritional conditions it was attained at 1140 g at 11 months of age. From a review of work on many species of deer Sadleir (1969) considers that puberty in deer is related to the level of nutrition available to the young during their prepubertal life but whether this involves the attainment of a particular body weight is not stated. Nutritional limitation of breeding is also seen in the red deer (*Cervus elaphus*) (Mitchell, 1973); hinds from the hill-land environment of Scotland annually produce 40 calves per 100 as a result of late puberty and long (greater than 1 year) inter-calving intervals, whereas in New Zealand 70 calves per 100 hinds are produced per year. Nutritional stress, particularly during pregnancy and lactation, appears to lead to the longer inter-calving interval in Scotland. Whether the weight/puberty relationship is generally true is questioned by Marshall (1978), who points out that genetic uniformity may contribute to the developmental uniformity observed in domestic species. Clearly more work is needed on this subject, similar to the studies of the wild rabbit by Myers and Poole (1962).

Circumstantial evidence for a link between weight and puberty or onset of breeding after anoestrus comes from a study of bank voles (*Clethrionomys glareolus*) and wood mice (*Apodemus sylvaticus*) where weights and the start of breeding have been observed over many winters with varied

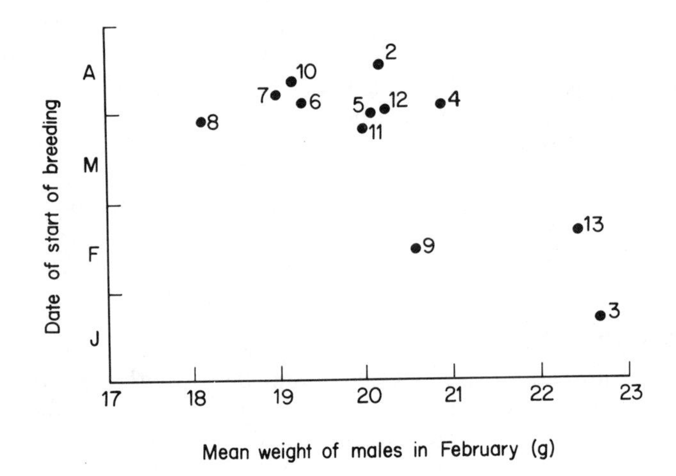

Fig. 3.11 Relationship between the mean weight of males in February and the onset of breeding (75% to 100% females with perforate vaginae) in wood mice at Wytham Woods, Oxford. Redrawn from Flowerdew (1973).

feeding conditions (Flowerdew, 1973). These small rodent populations in the wooded Wytham Estate near Oxford show some advance in the proportion of females manifesting vaginal opening and an increase in body weight compared with control populations when food supplies are artificially supplemented. Also, the body weights of male bank voles and wood mice (females may be heavier if pregnant and so confuse the issue) are generally heavier in the February of years when breeding (vaginal opening) started earlier (Figs. 3.11 and 3.12).

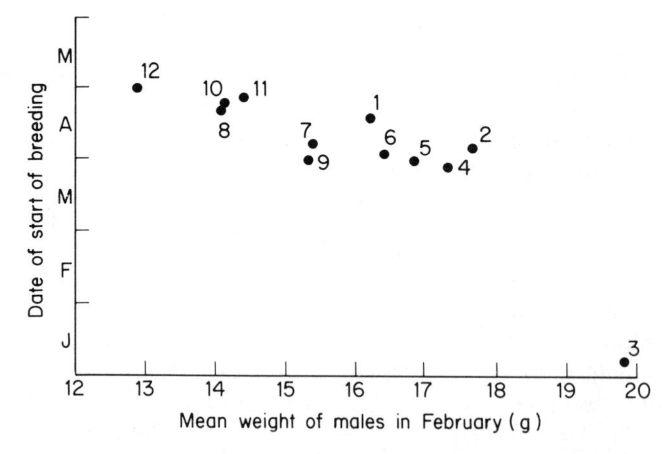

Fig. 3.12 Relationship between the mean weight of males in February and the onset of breeding (75% to 100% females with perforate vaginae) in bank voles at Wytham Woods, Oxford. Redrawn from Flowerdew (1973).

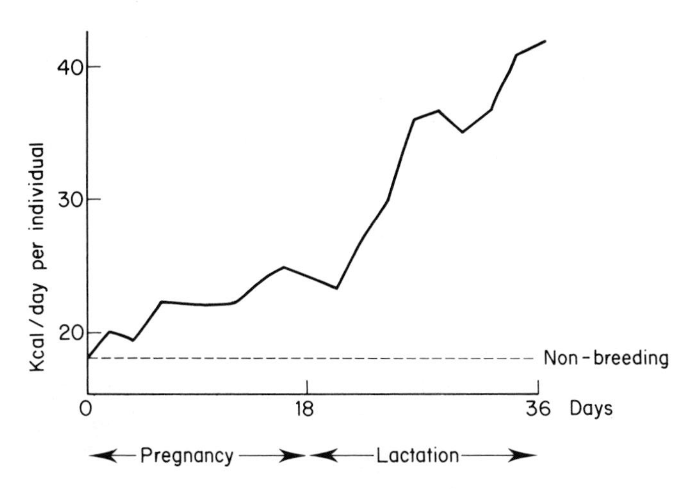

Fig. 3.13 Calorific intake (food assimilated) during pregnancy and lactation in the bank vole. The line represents the mean of 10 females. The mean calorific intake of non-reproductive females of the same size is shown by the dashed line. Redrawn after Kaczmarski (1966).

In sheep the well-known practice of 'flushing', increasing nutritional levels before mating, results in an increase in ovulation rate and a greater proportion of twin births; similar results have been observed from observations of white-tailed deer (*Odocoileus virginianus*) in contrasting habitats in North America (review by Sadleir, 1969). In domestic mammals decreases in nutritional levels can influence the interval between oestrous periods and cause the suppression of behavioural oestrus (in ewes) (Rattray, 1977). Food intake is generally increased during pregnancy and lactation; a striking example of the increase in calorific intake during these periods is shown by the bank vole (*Clethrionomys glareolus*) (Kaczmarski, 1966); Fig. 3.13 shows this phenomenon. The limitation of reproduction by available energy is shown by work on North American deer mice (Sadleir *et al.*, 1973) where estimates of the energy cost of lactation exceed estimates of energy consumed by females in winter (Fig. 3.14).

Undernutrition during pregnancy causes implantation failure, underweight young and other effects such as the total or selective mortality of young *in utero* or at birth. In the laboratory rat (*Rattus norvegicus*) reduction of food during pregnancy to starvation levels resulted in the resorption of embryos or still births (Russell, 1948). In mice such deaths of young are probably caused by a depression in the production or liberation of gonadotrophins (McClure, 1962). In Soay sheep, a breed which is free-living on the island of Hirta, St Kilda, single lambs born to heavier ewes have a better chance of survival than those born to ewes of a lower body weight (Grubb, 1974). Low birth-weight lambs have a poorer chance of survival but within each weight class of lambs there is a higher chance of survival with heavy ewes as mothers.

When the food supply of female rats is restricted during lactation, milk

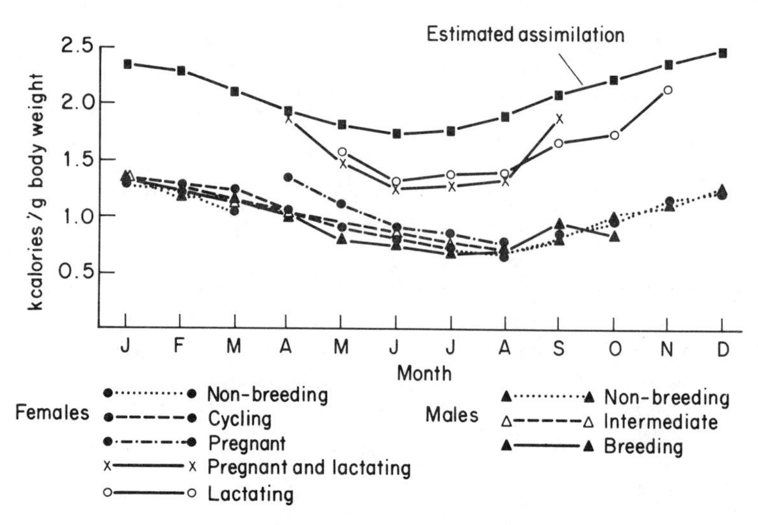

Fig. 3.14 Comparison of estimated assimilation and theoretical metabolic energy requirements of deer mice over a 24 hour period. Redrawn from Sadleir *et al.* (1973).

production takes place at a reduced level (Widdowson and Cowan, 1972). However, reduced milk production does not always follow food shortage, as female grey seals (*Halichoerus grypus*) starve during lactation and stored fat is utilized so that the young gain 20 kg in two weeks (Amoroso and Matthews, 1951). In woodrats (*Neotoma floridana*) when food is restricted during lactation the available energy is not always distributed evenly between the young (McClure, 1981); the 'investment' of lactation in young is altered when food is severely restricted so that there is a significant bias against males in both mortality and growth. It is common in bird reproduction to see parents favour one or more fledgelings over others when food is short.

Deficiencies in various nutrients can have marked effects on reproduction, but detailed accounts mainly come from domestic mammals (e.g. Rattray, 1977). Young male cattle deficient in protein, vitamin A and zinc suffer retarded sexual maturation; low manganese levels in gilts (female pigs) led to depressed behavioural patterns at oestrus or oestrus failure and Vitamin A deficiency led to irregular oestrous cycles. In the male domestic mammals Vitamin A deficiency is one of the most common causes of impaired reproductive capacity, leading to the inhibition of spermatogenesis and the depression of semen quality in a similar manner to the effects of undernutrition.

Some substances in food plants are capable of stimulating reproduction or inhibiting it. Studies of the montane vole (*Microtus montanus*) show that a plant-derived cyclic carbamate called 6-methoxybenzoxazolinone (6MBOA) extracted from wheat stimulates reproduction in free-living populations in winter when provided as a coating on oats (Sanders *et al.*, 1981); when reproductive activity was compared with a nearby population

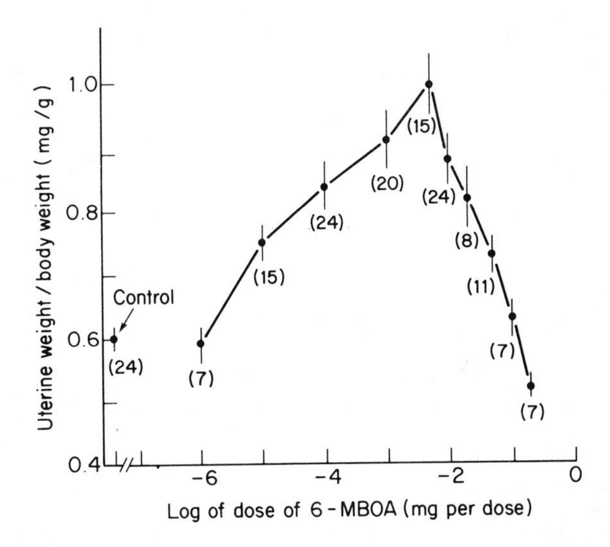

Fig. 3.15 Uterine weight of subadult montane voles (*Microtus montanus*) given varying doses of 6-MBOA. (Sample size in brackets.) See text for further details. Redrawn from Sanders *et al.* (1981). Copyright (1981) by the AAAS.

given only oats it was found that a high incidence of pregnancy and testicular hypertrophy occurred as a result of the treatment with 6MBOA (Berger *et al.*, 1981). Laboratory tests produced a dose-response curve for 6MBOA against uterine weight (Fig. 3.15). It is assumed that as 6MBOA is produced in growing shoots which have been injured it is possible that more 6MBOA is ingested when plant growth resumes after the winter. Thus in this facultative species reproduction would be triggered only when the vegetational conditions are improving and plant growth has begun, such as occurs after rainfall (Negus *et al.*, 1977); drought causes the cessation of reproduction and the resorption of embryos. It is also suggested (Berger *et al.*, 1977) that reproduction in this species may be prevented, once the stimulating substance has stopped being produced, by phenolic compounds in the grass. It is interesting to note that 6MBOA is also stimulatory to the ovarian weight of the laboratory mouse (Sanders *et al.*, 1981). How general the stimulation and inhibition of reproduction by plant products may be in mammals is yet to be ascertained; oestrogens and 'plant hormones' are common in vegetation (Bradbury and White, 1954) and their importance is still a matter for study.

Behavioural and social effects on reproductive activity

The influence of other individuals on the reproductive activity of a male or female mammal is obvious during competition for mates (see Chapter 2). However, more subtle chemical, auditory (including high frequency) or tactile influences may be at work alongside the overt aggression or agonistic behaviour patterns which are usually observed.

Odours and other chemical factors affecting reproduction

Odours and less volatile chemical cues emanating from specialized scent glands, excretory products or faeces impinge upon the olfactory and accessory olfactory (vomeronasal) systems providing information about other individuals and also modifying the recipient's behaviour or physiology. The study of scent glands and their role in mammalian communication is a rapidly developing area; reviews of scent marking are provided by Ralls (1971) and Johnson (1973) and broad reviews as well as specialized accounts are provided by Aron (1979), Muller-Schwarze and Silverstein (1980), Stoddart (1980), Marchlewska-Koj (1984) and Brown and Macdonald (1985), among many others. Many mammalian species mark their home range or territory with urine and other scent marks and if these are detected by other individuals of the same species then behavioural or reproductive changes may result.

Olfactory cues are generally divided into 'signalling' or 'releasing' odours which result in an almost immediate change in motor activity or they can be priming odours which affect neuroendocrine and endocrine activity (Wilson and Bossert, 1963; Bronson, 1974); odours may also act to aid recognition during development and thus be imprinted by individuals. A response to a signalling odour such as a sex attractant will be immediate and made by a change in behaviour whereas a priming factor such as house mouse urine will induce neural and hormonal changes in the recipient female which will result in ovulation and sexual behaviour 48–72 hours later. It is likely that odours also convey information from one individual to another and that species, family group, age or social state of maturity, sex, sexual state (in cyclic females) and individual identification are transmitted in this way; odours may also induce aggression or 'fear' as well as helping in the maternal recognition of offspring and vice versa. Such odours transmit information or act as cues for behavioural or physiological events and in the latter two cases parallel insect 'pheromones'. The definition of a pheromone is, strictly speaking, an odorous substance which has a specific response in the same species and it is usually a single substance but may be a mixture. Odour cue is perhaps a better term to use in the mammalian context as these substances may be more general in their action and impart information to other species; also the response may not be as stereotyped as many insect responses apparently are. Mammalian odour cues may elicit a reaction which depends upon past experience and may be variable in nature; early experience of odours may determine the adult's response and the stimulus may involve other senses such as sight, sound or touch.

Priming odours

Priming olfactory cues in mammals act on developmental and physiological processes such as the expression of oestrus, puberty, testicular development and the success of pregnancy, as well as the associated neuroendocrine and endocrine systems. One caveat is necessary: it is possible that some observed priming effects are laboratory artefacts and serve no purpose in wild populations.

Central to the theme of reproductive priming is the regulation of the timing of oestrus and ovulation. The study of odour cues which affect reproduction owes much to the work of Whitten (1956, 1958, 1959) who kept female laboratory mice in groups of 30 and observed that the frequency of oestrous vaginal smears from these mice was retarded compared with those kept in single cages because of prolonged dioestrus. This mutual suppression of oestrus has been labelled the 'Lee-Boot effect', after the work of Lee and Boot (1955, 1956) who observed a high incidence of long oestrous cycles, corresponding to pseudopregnancies, in mice kept four to a cage compared with mice kept singly. The same effects are observed if females are kept alone in cages recently soiled by groups of females of the same strain (Champlin, 1971) and thus an olfactory cue was likely to have caused the effect. Suppression of oestrus in grouped females is even seen when the optic nerves are sectioned, ear drums ruptured and tactile sensation (contact) prevented by subdivision of the cage (Whitten, 1959).

Mutual suppression of oestrus by grouping females may be modified by the addition of a male mouse. The modification takes the form of synchronization of oestrus, most vaginal plugs being found on the third night after the introduction of the male stimulus (Whitten, 1958, 1966). Odour is again implicated as male urine and a caged male both synchronize oestrus (Fig. 3.16) and similar effects are produced if only 100 μl of male urine per

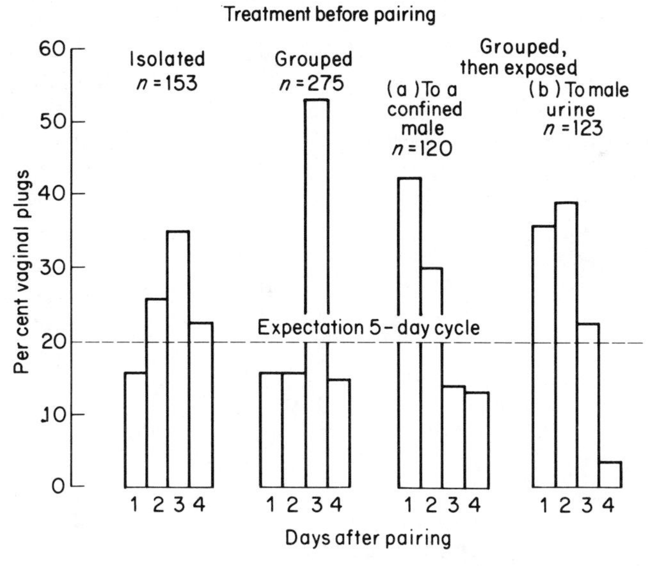

Fig. 3.16 Frequency distribution of vaginal plugs (present following copulation) in female domestic mice after pairing. The mice had been exposed to four previous treatments: isolated, grouped, grouped and exposed to confined males and grouped and exposed to male urine. The histograms show the percentages of the total number of vaginal plugs occurring within four days. Synchronization of oestrus is best shown by females which were grouped before pairing. Redrawn from Whitten (1966). © Grafton Books – a division of the Collins Publishing Group.

Table 3.1 Mean age ($\pm$ S.E.) of female house mice at vaginal opening and at first oestrus, after various lengths of prepubertal exposure to an adult male (from Vandenbergh, 1967). © The Endocrine Society.

Age of females at introduction of male	Age (days) at vaginal opening	Age (days) at first oestrus
21 days	30.5 $\pm$ 0.88	37.1 $\pm$ 1.62
30 days	30.4 $\pm$ 0.83	41.9 $\pm$ 1.00
38 days	32.9 $\pm$ 0.77	45.6 $\pm$ 1.03
No male	34.9 $\pm$ 1.24	57.1 $\pm$ 1.83

day is applied to the bedding for 2 days (Marsden and Bronson, 1964). Thus male urine apparently overcomes any inhibitory effect of female odours and stimulates the expression of oestrus, either by earlier follicle development or by the increased growth rate of follicles. The synchronization of oestrus by male odour is known at the 'Whitten effect'.

The induction of oestrus at puberty is also accelerated by olfactory influences from the male. Immature female house mice reared in the presence of an adult male from 21 days of age showed first oestrus 20 days before females reared in isolation (Table 3.1) (Vandenbergh, 1973). Puberty may even be induced by placing male urine on the nose of the young female (Colby and Vandenbergh, 1974). It seems that the active substance in male urine has been partially isolated (Vandenbergh *et al.*, 1975; Vandenbergh *et al.*, 1976); it is thought to be a protein of about 860 molecular weight or else linked to such a protein and it is non-volatile. Results of experiments inducing oestrus where males are separated from females are explained by the movement of the substance on soiled bedding. However, Bronson (1979) suggests that as the stimulus seems to be capable of being carried downwind in a controlled air chamber it may be carried as an aerosol, or else the active fraction involved is small and is a potentially volatile substance which is bound to a larger protein. It is puzzling that another paper by Monder *et al.* (1978) suggests that the primer involved is a lipid.

The action of the puberty-accelerating odour is enhanced by tactile cues, possibly fighting and chasing, from the male (Bronson and Maruniak, 1975); females having occluded eyes and ears kept with castrated males, as well as urine from intact (non-castrated) males, showed enhanced uterine growth similar to that shown when exposed to a castrated male and intact male urine without the sensory deprivation.

The urinary odour cues inducing oestrus are dependent upon the presence of testosterone, and presumably act via the nervous nuclei in the brain that are associated with the hypothalamus and the anterior pituitary gland to affect FSH and LH production, and hence follicular growth and ovulation (see Chapter 1). It is uncertain whether the cue is specific to one species, as different workers have found conflicting results on this point (Bronson, 1979). It may be presumed that the active substance is similar or identical to that which induces oestrus in the 'Whitten effect' (Bronson, 1974). Similar stimulatory effects of urine on female puberty have been found in other rodent species such as the prairie vole (*Microtus ochrogaster*) (Carter *et al.*, 1980), the white-footed mouse (*Peromyscus leucopus*) (Rogers and Beauchamp, 1976), the prairie deermouse (*Peromyscus maniculatus bairdii*) (Teague and Bradley, 1978) and in rats (*Rattus norvegicus*) (Vandenbergh, 1976).

The synchronization, suppression and advancement of oestrus may occur by olfactory means in other species of mammals. The Lee-Boot and Whitten effects also occur in the prairie deermouse. In non-rodent species the synchronization and shortening of the oestrous cycle occurs in ewes on the introduction of a ram to the flock (Thibault *et al.*, 1966) and olfactory cues from the ram are implicated in the stimulation of ovulation (Knight and Lynch, 1980). Reduced oestrous cycle duration is thought to occur in the goat under the influence of the male but the role of olfaction is not fully established (Aron, 1979); cervical mucus from oestrous cows acts as a primary odour helping to synchronize oestrus in heifers (Izard and Vandenbergh, 1982).

In free-living populations either the advancement of puberty or the synchronization of oestrus could occur. It is possible that a dispersing female might meet a male and, if immature, reach first oestrus earlier than if the male was absent from the area; presumably under 'normal' conditions, however, the male influence is always present. The latter applies to the synchronization of oestrus and it would be unusual for many females to be in a non-pregnant state in typical rodent populations where post-partum oestrus is common.

A further laboratory olfactory effect is that females will inhibit the puberty of other females. There is a positive linear relationship between the number of female mice per cage and the mean age of vaginal opening (Castro, 1967; Vandenbergh, 1973) and even soiled bedding from groups of females will inhibit maturation of an isolated female (Drickamer, 1974; Colby and Vandenbergh, 1974). Similar effects of male presence on the testicular development of young male mice have also been described but the effect apparently lasted only up to 78 days of age and the relative importance of physical stimuli or olfactory stimuli was not elucidated (Vandenbergh, 1971).

Perhaps the most famous olfactory influence on mouse reproduction is the 'Bruce effect' or pregnancy block which occurs in the house mouse in the early days of pregnancy if the newly mated female is introduced to a strange male (Bruce, 1959). This occurs in non-post-partum females on removal from the stud male and normal oestrous cycles resume within a short time; the effect also occurs if the male is caged within the female's cage to prevent physical contact but anosmic females will not react to strange males in this way. The length of exposure needed to obtain a maximum response is 3 days and the female remains susceptible up to 5 days after mating, i.e. until implantation. Pregnancy block also occurs in deer mice (*Peromyscus maniculatus*) (Bronson *et al.*, 1969), prairie voles (*Microtus ochrogaster*) where the presence of a strange male stops implantation or causes abortion up to day 15–17 of pregnancy (Stehn and Richmond, 1975), and probably in other *Microtus* species. Evidence that pregnancy blocking occurs in the wild comes from studies of meadow voles (*Microtus pennsylvanicus*) where it seems that females in their first pregnancy are susceptible to blockage but nursing females are not (Mallory and Clulow, 1977). However, it has not yet been shown to occur in wild house mice (*Mus musculus*) (Bronson, 1979); it is argued that the blocking mechanism is the same as that used by males to induce ovulation in non-pregnant females and that none of the arguments for the adaptive significance of blocking are convincing. It is also difficult to visualize the occurrence in the wild of the specific social conditions necessary to produce the block. Oestrus suppression is also difficult to visualize in the wild as females would probably find it difficult to avoid the stimulatory male odours.

Signalling odours

Signalling odours in mammalian reproduction are involved in the attraction of the two sexes, in mother–young relationships and in eliciting mating behaviour as well as imparting information about reproductive condition. As many mammals show scent-marking behaviour using urine, faeces or body secretions (Ralls, 1971; Johnson, 1973) it is not surprising that the odours from individuals act as attractants prior to mating. These odour cues may act alone or together with tactile, visual and auditory signals.

In house mice olfactory cues form an important mediator of sensory information between individuals which are often nocturnal and remote in space and time. Females become more active in the presence of odour from male cages (Ropartz, 1968) and are more attracted to urine from intact males than to urine from castrated males (Scott and Pfaff, 1970). It seems that the preputial glands (see Chapter 1) may be important in the production of a female-attracting substance in mice (Bronson and Caroom, 1971) and that the substance is deposited during urine-marking. It is much more attractive to sexually experienced females than to sexually inexperienced females. In contrast, female laboratory rats are attracted to the odour of males whether they (the females) are sexually experienced or not (Carr, *et al.*, 1965). Such phenomena as sex-attraction must be common in many mammals and there is evidence to indicate that it occurs in such species as the dog, cat, ewe and sow where the female is drawn to the odour of the male although other senses may also be involved (Aron, 1979).

In the attraction between the sow and boar the sow is drawn to the boar at a distance by an as yet unidentified odour, and when close attraction is by a metabolite of male hormones (androgens) called 3αhydroxy-5αdrost-16-one produced by the salivary glands. Another androgen metabolite (5αandrost-16-en-3-one found in the preputial sac) and other similar steroids elicit the standing reaction in oestrous sows (lordosis) which is necessary before mating can take place (Signoret, 1976). Aerosals containing this substance are sold commercially as 'Boar Mate' (Antec A.H. International Ltd). In the attraction of males to females such evidence has accumulated to implicate the role of female urine or body secretions in the attraction of males to females in a number of species. In rodents, adult male rats prefer the odour from receptive females to that of non-receptive females (Le Magen, 1952), and in mice female urine increases the activity of males and induces the production of ultrasonic calls in socially experienced males (Smith, 1981). However, the most attractive odour in mice, as far as the indication of receptive reproductive condition is concerned, is that of the vaginal secretions (Hyashi and Kimura, 1974); males are found to mount more with dioestrous females smeared with oestrous female vaginal secretions than with those smeared with oestrous female urine. Only sexually experienced male mice show a preference for oestrous females whereas naive males show no preference for secretions from oestrous females over those from mice in anoestrus or dioestrus. In the golden hamster (*Mesocricetus auratus*) vaginal secretions are attractive to males and males show a preference for vaginal odours even if they are sexually inexperienced males (Gregory *et al.*, 1975). Castration reduced this preference and treatment with testosterone propionate partially restored it, thus the hormonal status of the male affects the preferences more than sexual experience.

In dogs, males prefer oestrous female urine to dioestrous urine (Doty and Dunbar, 1974), and can detect a vaginal secretion in oestrous females which stimulates sexual behaviour and mounting attempts; the substance inducing this behaviour is methyl p-hydroxybenzoate (Goodwin *et al.*, 1979). Studies of the rhesus monkey (*Macaca mulatta*) suggest that vaginal secretions are attractive to males (Michael *et al.*, 1971; Curtis *et al.*, 1971) and that the attractive substances are short-chain fatty acids, however, later work (Goldfoot *et al.*, 1976) has cast doubt on this conclusion. In sheep the odour of females is attractive to males but visual and auditory cues are also important, especially when the partners are in close proximity. (Fletcher and Lindsay, 1968; Lindsay, 1965); other ungulates such as bulls, however, appear not to need olfactory cues to become attracted to the female as immobilization of the cow is sufficient stimulus to induce mounting by the bull (Aron, 1979).

Urine testing, 'flehmen' and the vomeronasal organ

Most mammals possess two 'olfactory' systems: the nervous pathways to the olfactory areas of the brain from the nasal cavity and the accessory olfactory system from the vomeronasal organ, or organ of Jacobson, to the accessory olfactory bulbs or equivalent areas of the brain. Jacobson's organ is found in every order of mammals except possibly the Cetacea; it consists of bilateral, elongated, tubular structures flattened from side to side and located medially in or on the anterior parts of the nasal septum (Wysocki, 1979). In eutherians the organ is best developed in lagomorphs and rodents and is relatively larger in small than in large species (Estes, 1972). The position of the organ's opening is variable; in rodents it opens into the nasal cavity (Fig. 3.17), in cats it runs into the nasopalatine canal, a small bilateral passage through the anterior of the palate to the nasal cavity and in other species, for example the cow and hippo, it opens directly into the oral cavity.

The function of the vomeronasal organ has been the subject of much

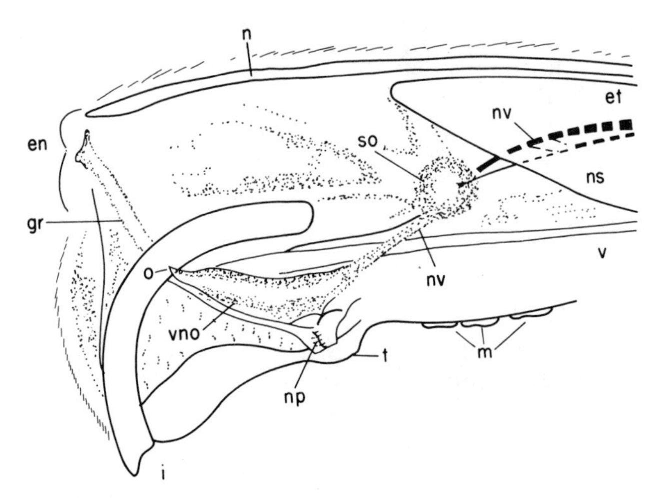

Fig. 3.17 Diagram of the lateral view of the nose of a Guinea pig showing the vomeronasal and septal organs and their relationship with the olfactory region (triangular area on the right). Abbreviations: en = external naris; et = endoterbinates; gr = groove formed by stratified squamous epithelium; i = incisor; m = molar; n = nasal bone; np = nasalpalatine duct; ns = septal organ nerve bundle; nv = vomeronasal nerve bundle; o = opening of vomeronasal organ; so = septal organ of masera; t = midline tubercle (palatine papilla); v = vomer; vno = vomeronasal organ.

controversy in the past but recent evidence supports the idea that it is a chemosensory structure (Wysocki, 1979) and Estes' (1972) hypothesis that the function is urinalysis has carried support. Studies of the Guinea pig (Wysocki *et al.*, 1980), where males presented with urine mixed with fluorescent dye showed dye-specific fluorescence in the vomeronasal organ, indicate such a function is possible. The route by which the dye reaches the vomeronasal organ has been postulated to be a result of 'flehmen', the distinctive facial grimace where the lips are curled back and the nares closed, and which is associated in male mammals with the presence of female urine (Estes, 1972). The grimace is commonly seen in many ungulates such as red deer during rutting, domestic mammals, cats, bats and hyaenas. It is possible that urine could be carried into the vomeronasal organ on air currents resulting from differential air pressures in the body of the organ and in the nasal or oral cavity (Estes, 1972), but Wysocki (1979) prefers the idea of a pumping action which is not necessarily dependent upon flehmen but which could function with this act. Work on the goat (Ladewig and Hart, 1980), where individuals are encouraged to perform flehmen in response to female urine, indicates that only in male goats performing flehmen did fluorescent die mixed with the urine reach the posterior sensory part of the vomeronasal organ. Flehmen behaviour in black-tailed deer (*Odocoileus hemionus columbianus*) commonly involves mouth and tongue contact with urine and the initial interest of the male in female urine varies with season, increasing during the rut but stopping or becoming rare prior to copulation, indicating that flehmen is probably involved in establishing the state of oestrus of the female (Muller-Schwarze, 1979). Once behavioural oestrus is manifested by the female and she stands for copulation then flehmen is no longer necessary.

Urine analysis seems to be one function of the vomeronasal organ; however, experiments where the olfactory and vomeronasal inputs to the brain are removed show disruption of reproductive social behaviour in many mammals; more precise studies of male hamsters where intranasal zinc sulphate infusion destroyed the nasal epithelium but not the vomeronasal epithelium, show that reproductive behaviour generally persists (Powers and Winans, 1975; Winans and Powers, 1977). Thus the mediation of stimuli eliciting sexual behaviour can be added to urine testing as a function of the vomeronasal organ. A further function appears to be the involvement with odour stimuli suppressing oestrus in grouped mice (Reynolds and Keverne, 1979). In a study of rats, Johns *et al.* (1978) found that many female rats in light-induced persistent oestrus ovulated in response to male urine; however, when physical contact with the odorized substrate was prevented ovulation did not occur; when the vomeronasal openings were blocked and the females were presented with male-soiled bedding only 2 out of 22 studied showed ovulation. Further research may implicate the vomeronasal organ in many other chemosensory responses but it seems likely that urine analysis is one of the common functions, particularly in ungulates.

Maternal odours and maternal relationships

The difficulty in cross-fostering young lambs when the ewe's lamb has died is common knowledge. Shepherds place the skin of the dead lamb on the orphan to deceive the ewe into thinking it is her own. Such olfactory influences on maternal behaviour are common in mammals and have been particularly studied in ungulates and rodents.

The imprinting of the odour of the young on the female forms the basis of the maternal relationship and can occur in a very short time. In domestic goats the first five minutes of contact between mother and young are enough to induce rejection of all other young; anosmic females will accept any young (Klopfer and Gamble, 1966). Rodents are not so choosy and accept other females' young even from another species (McCarty and Southwick, 1977). Olfactory attraction also operates between the mother and young; in rats the females produce an odour which is attractive to young from about days 14–27 of lactation (Leon and Moltz, 1972; Leidahl and Moltz, 1977). The attractive odour is produced by the caecum and is passed out with the faeces as well as tainting the whole body. It also occurs in spiny mice (Porter and Doane, 1976; Porter *et al.*, 1977) and probably in many other species; the attraction of the odour to the young is not species specific and the strength of attraction depends upon the diet of the lactating female. Odour from another species kept on the same diet as the lactating spiny mouse was more attractive than odour from the mother when fed on a novel diet.

Conclusion

The environment acts in diverse ways on many physiological and behavioural processes associated with reproduction. Different components may act in concert or singly, and where much progress is required is in the assessment of environmental effects on reproduction in the field. Controlled studies in the laboratory can only tell us so much about animal/environment interactions but are nevertheless necessary to assess the likely importance of various factors in the field.

Temperature, photoperiod and nutritional factors affecting reproduction are relatively direct and easy to observe in comparison with social effects. These latter may not be immediately obvious without special knowledge of the communications systems operating between individuals, be they visual, olfactory, auditory or other.

4

Interactions Between Reproductive Biology and Life-history Parameters

Some species breed faster than others. This is not only true of species from different ends of the animal kingdom but also true of species within a class such as the mammals. Even the same species may show a variation in reproductive output throughout its range. This chapter demonstrates some of the variation in reproduction and explains why it should occur. Such explanations are not always obvious and much of the subject is couched in theory which has yet to be confirmed in the field. However, many of the theories are quite feasible and at least give an insight into the possible causes of particular events.

Much of the variation in reproductive parameters can be explained by the size of the mammal. The common measure of reproductive rate, the intrinsic rate of natural increase, r, (see Chapter 5) is defined as $\log R_0/T$, where R_0 is the net reproductive rate and T is the mean generation time. The logarithm of this measure, r, decreases with the logarithm of body weight for a large range of animal species (Fenchel, 1974), and thus it might be expected that many, if not all, reproductive parameters would also vary with body weight. This possibility will be discussed in this chapter. Other variables than body weight may be important in influencing reproduction; parental care, protection from predators, availability of the food supply may all play a part and the importance of these factors can be investigated in theory as well as in practice.

Evidence for the adaptation of reproduction to the environment

Latitudinal and altitudinal variation in litter size

Comparative studies of the same or closely related species from different geographical locations have produced interesting results, particularly when litter size is studied. Lord (1960) obtained litter size data for closely related species of mammals from North America and came to the conclusion that many non-hibernating prey species showed a positive relationship of litter size to latitude (Fig. 4.1). However, the hibernating ground squirrels (*Spermophilus* species) and the fossorial gophers (*Geomys* and *Thomomys* species), as well as many predatory species (particularly the mustelids and the cats) did not show this relationship. A study of deer mice (*Peromyscus maniculatus*) from California (Dunmire, 1960) also showed that litter size increased with altitude and this was supported by further studies on the

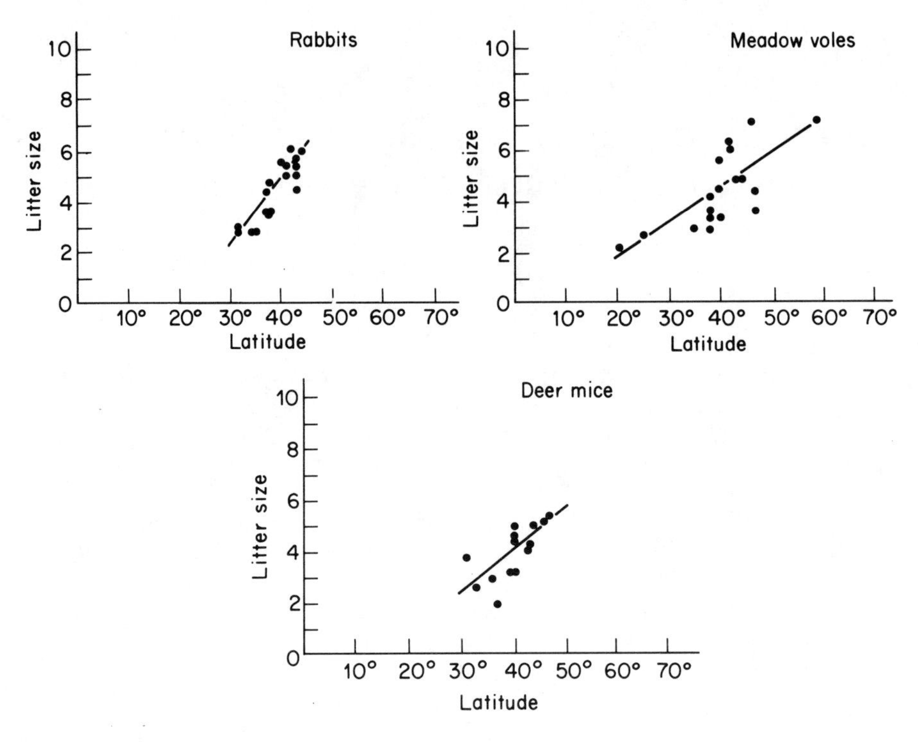

Fig. 4.1 Examples of mammalian taxonomic groups which show a significant positive correlation of litter size with latitude. Redrawn from Lord (1960).

same species in Colorado (Spencer and Steinhoff, 1968).

Variations such as these have attracted a number of explanations. Lack (1947–48) suggested that the increase in clutch size of birds with latitude was the result of the birds' ability to find more food for their young as day length became greater with latitude. As Lord (1960) points out, this is hardly applicable to mammals as many species will be nocturnal and be forced to seek food in daylight during the long arctic summer days. However, Lack's more generally accepted argument, that the modal or mean number represents the most consistently successful clutch or litter, has been developed by Spencer and Steinhoff (1968). They suggest that the shorter seasons of the more northerly latitudes or higher altitudes limit the number of times an animal resident in those areas is able to reproduce in its lifetime compared to its relatives at lower latitudes or altitudes. It therefore becomes advantageous for the female to invest its energies in a few, large, early litters, even though doing so may reduce the life expectancy of the female. Short seasons make it impossible for the female to reach the normal maximum reproductive effort attainable with smaller, more frequent litters in an environment which permits a longer breeding season. Particularly with the smaller species where longevity is likely to be a few breeding seasons at the most, the possible reduction in longevity caused by larger litter sizes would be more than compensated for by the increased number of young produced. Species which do not show a latitudinal variation in litter size, such as the fossorial and hibernating species mentioned above, often have only one litter annually and so the effect of shortening the breeding season would have little influence on their reproductive success in terms of the number of surviving offspring. This latter explanation seems quite plausible but other theories may be equally acceptable; Fleming and Rauscher (1978) suggest that the altitudinal variation in litter size may be the result of a skewed age distribution so that older females with larger mean litter sizes dominate the populations at higher elevations. This would imply that the variation does not have a genetic basis, but this is certainly present in snowshoe hares (Keith *et al.*, 1966).

From the evidence of variation and the arguments set out above it does seem that litter size may be an adaptive feature of reproduction in many species of mammal. How general a feature this is and which other aspects of reproduction are strongly adaptive are discussed in the next section.

Relationships between adult body weight and reproductive variables

In comparisons between species of mammals it might be expected that if a particular reproductive variable does not show a strong relationship with adult body weight then it is likely that the variable has become adapted to the environment. Rate, shape and size characteristics tend to vary in a non-linear fashion with the total size of the animal and these allometric relationships are described by an equation of the form

$$Y = am^b$$

where Y is a variable, m is adult body mass in grams and a and b are derived empirically by fitting a least squares regression line to a series of points. These equations have a linear form

$$\log Y = \log a + b\log m$$

Millar (1977) used these assumptions and formulae to examine the allometric relationships of reproductive and developmental characteristics to the adult body mass of mammals. Despite a lack of larger mammals in the data (100 species, mainly rodents, lagomorphs, soricids, tenrecs, bats, a procyonid and a cervid; no aquatic mammals and lacking many large herbivores, carnivores and primates) they raise some interesting points regarding the adaptation of reproductive variables.

Relative birth weight varies from less than 4% of adult weight (many rodents, lagomorphs, the sciurids, a tenrec and the procyonid) to about 20% of adult weight (some bats). This is clearly related to adult body size as the smallest mammals (the bats) have the largest relative birth weights. However, birth weight increases as the 0.71 power of adult mass in 95 species studied. This relationship is considered to be highly predictable but some groups do show constant deviation from the expected relationship. The 17 sciurid species studied have birth weights less than those predicted from the general equation, which is dominated by murid rodents and lagomorphs, and bats have birth weights heavier than expected. This variation is illustrated in Fig. 2.12, where hystricomorph rodents show an increase in log birth weight as the 0.86 power of log adult weight, which is greater than the power of the myomorph rodents at 0.75 and the sciuromorphs, as stated above, below the norm at 0.62 (from 16 species examined by Case, 1978). High birth weights seem to be related to low litter sizes and vice versa.

Weaning weight and growth rate to weaning are both strongly related to adult weight with no consistent deviations from the general equations within taxonomic groups. Case (1978) extended this analysis and concluded that litter size explained little of the variation in mammalian postnatal growth rates but that birth weight usually has a more significant and positive relationship with growth rate. Case suggested that given a fixed amount of energy for reproduction, a larger birth weight could be accommodated either by decreasing the litter size or by extending the gestation period (or both). Alterations in the rate of embryonic growth appear to be severely constrained by the rate at which brain tissue grows and differentiates (Sacher and Staffeldt, 1974). The fact that birth weight and postnatal growth rate seem to be related may be explained, as Case suggests, if the selective factors influencing relative birth weight and postnatal growth rate are similar. Stearns (1976) suggests that a large offspring is favoured when infant mortality rates are high and when resources are abundant and continuously distributed. Case shows that these appear to be the two basic demographic and environmental features associated with rapid postnatal growth in many mammals. Table 4.1 shows the results of compiling infant mortality rates (up to reproductive maturity) and their relationships to adult mortality rates. Mortality is expressed as the percentage of deaths per period of time given in the first column (some rates are converted from one time base to another, see Case (1978) for details). These results generally support the above argument and the predictions of Williams (1966), Gadgil and Bossert (1970) and Case (1978), that the primary sources of infant mortality may be different in fast-growing and slow-growing animals. Species with slow postnatal growth presumably avoid mortality which fast growers would incur. Slow growers would be subject to starvation, small nest predators and adverse physical conditions. Fast growers would be subject to accidents, diseases, bad weather or large nest predators which can overpower the parents. In looking for examples the fast growing artiodactyls may be compared with the slow growing elephants. The former suffer mortality primarily from predation and secondarily from drought-related starvation in many cases, whereas the latter suffer little predation. Young lagomorphs often have high mortality rates as a result of predation, disease and environmental exposure; they show a fast postnatal growth rate. Although there are other species which do not fit the predictions or for which information is lacking, further corroboration is found from many species which show a higher growth rate in juvenile males than in juvenile females and correspondingly show a higher mortality rate in juvenile males than females.

Litter size in mammals varies from one (as in many ungulates) to more

Table 4.1 Mammalian infant mortality according to relative growth rate categories (from Case, 1973) with modifications to nomenclature (nd = not determined).

Species	Time units	Infant mortality (%)	Mortality ratio infant/adult
FAST GROWERS			
Lepus americanus, snowshoe hare	month	75	2.8
L. americanus	month	24*	4.0
Ochotona princeps, pika	month	28*	3.3
Canis lupus, wolf	year	57–94	2.6–4.3
Vulpes cinereoargenteus, grey fox	8 mth	65	1.4
Panthera leo, lion	year	36–67	6.0–11.0
Phoca hispida, ringed seal	month	35	17.5
Monachus schauinslandi, Hawaiian monk seal	month	24	50.0
Halichoerus grypus, grey seal	month	19	12.0
Mirounga angustirostris, northern elephant seal	month	8.7–14	nd
Equus burchelli, zebra	year	19	1.6
Syncerus caffer, African buffalo	year	49	8.0
Hemitragus jemlahicus, Himalayan thar	year	53	3.7
Rangifer tarandus, reindeer	year	40–75	6.4–14
Saiga tatarica, saiga	year	70–90	2.3–3
Giraffa camelopardalis, giraffe	year	73	24.0
Odocoileus hemionus, black-tailed deer (M)	year	53–72	2.5–3.6
(F)	year	37–53	1.9–2.7
Ovis dalli, Dall sheep	year	72	9.0
MODERATE GROWERS			
Clethrionomys glareolus, bank vole	month	39	1.3
Apodemus flavicollis, yellow-necked mouse	month	27	1.2
Proechimys semispinosus, Tomes's spiny rat	month	18	1.1
Oryzomys capito, rice rat	month	28	0.9
Liomys adspersus, Panamanian spring pocket mouse	month	6.1	0.8
Microcavia australis, cavy	month	36	1.8
Eptesicus fuscus, big brown bat	month	8*	3.5
Tadarida brasiliensis, Brazilian free-tailed bat	month	10*	nd
Myotis thysanodes, fringed myotis bat	month	1.3*	nd
M. lucifugus, little brown myotis bat	month	1.9*	nd
Mephitis mephitis, striped skunk	year	75	0.9
Callorhinus ursinus, northern fur seal	month	2.4–3.6*	nd
Phacochoerus aethiopicus, warthog	year	50	1.5
SLOW GROWERS			
Megaleia rufa, red kangaroo	year	39	2.6
Sarcophilus harrisi, Tasmanian devil	month	0	0
Trichosurus vulpecula, brush-tailed possum	year	15	0.6
Ursus americana, American black bear	year	5	0.4
Loxodonta africana, African elephant	year	12	2.4
Macaca mulatta, rhesus monkey	year	18	2.0
Macaca mulatta	year	18	3.6

*Mortality rates converted from the original time units given by the investigator.

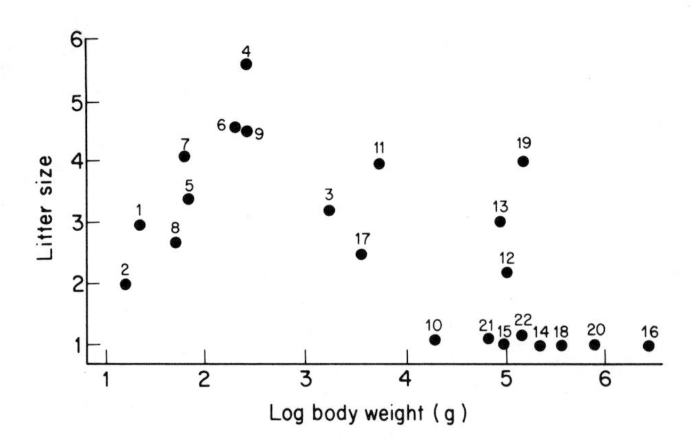

Fig. 4.2 The relationship between litter size and body weight in a range of mammalian groups. The following figures in parentheses indicate the number of species 1 Soricidae (2), 2 Vespertilionidae (3), 3 Lagomorpha (8), 4 Sciuridae (15), 5 Cricetinae (30), 6 Microtinae (10), 7 Gerbillinae (3), 8 Heteromyidae (6), 9 Muridae (4), 10 Primates (21), 11 Canidae (4), 12 Ursidae (5), 13 Felidae (5), 14 Pinnipedia (4), 15 Delphinidae (3), 16 Proboscidea (1), 17 Hyracoidea (1), 18 Perissodactyla (2), 19 Suidae (1), 20 Hippopotamidae (2), 21 Tylopoda (2), and 22 Ruminantia (15). Data of groups 1–9 from Millar (1977) and 10–22 from Sacher and Staffeldt (1974). Redrawn from Tuomi (1980).

than eight (e.g. domestic pig and rat). In general, sciurids, canids, murids, cricetids, soricids and lagomorphs have high litter sizes with means around five and six, whereas the vespertilionids, phocids, most artiodactyls and cetaceans have lower litter sizes with means just above one. Millar's (1977) analysis where litter size is related to adult body weight gives a non-significant result whereas Sacher and Staffeldt (1974) obtain a negative slope and therefore b in the allometric equation is negative. Thus it seems that the sample of mammals taken for the analysis might have an effect on the form of the relationship and this is shown very well by Tuomi (1980), who realized that the relationship between litter size and adult body weight changes as body size increases (Fig. 4.2). Thus up to 1 kg adult weight, litter size increases with body weight and at weights above this the litter size generally decreases, but there is much variation, indicating possible adaptation. The smaller species groups have smaller litter sizes than medium-sized mammals because of their relatively high birth weight, there is probably a minimum birth weight below which mammals do not go unless special arrangements are made for parental care, as in the marsupials. Large mammals have small litters because of their relatively small amount of resources invested in reproduction and the high absolute birth weight. Mammals of intermediate size have a relatively low birth weight and invest a large amount of resources into reproduction and are therefore able to have large litters. Large mammals may be able to increase litter size by decreasing birth weight but small mammals cannot do this, as noted above. However, small mammals may increase litter size by increasing reproduc-

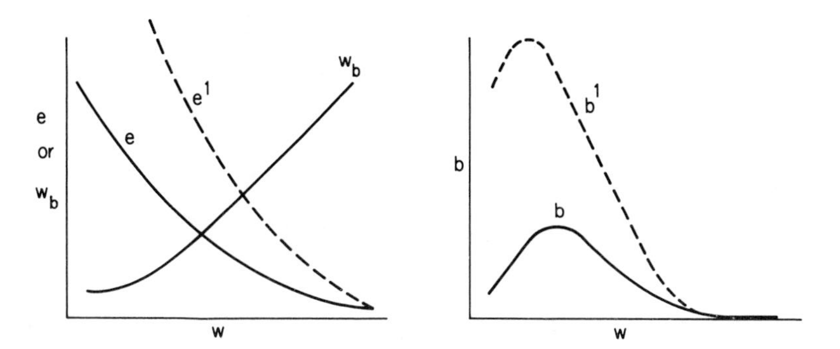

Fig. 4.3 The relationship between litter size, b, and body weight, w (*right*) predicted on the basis of relative litter weight, e, and birth weight, w_b (*left*). e^1 and b^1 represent an hypothetical situation where resource availability is unusually high. Redrawn from Tuomi (1980).

tive effort (energy/resources invested in reproduction) when the food supply is available and this would lead to a simple negative relationship between litter size and adult body weight (Fig. 4.3).

The evolution of reproductive parameters

Litter size has obviously become an adaptive variable in mammalian reproduction but few studies have shown why a particular litter size might have evolved. An experimental study which attempts to answer this question has been carried out by Fleming and Rauscher (1978). They ask why the litter size of the white-footed mouse (*Peromyscus leucopus*) is on average close to 5. Laboratory studies showed that the mean litter size for the first litter was 4.1, for the second litter 5.3 and for the third and later litters 5.0. Litter size varied significantly with age and parity (Fig. 4.4) and it is concluded that more reproductively experienced females gave birth to litters which survived better than those of younger females. Thus a litter of 4 is most productive in reproductively inexperienced females but larger litter sizes are more productive still in older females. Curiously the weight at birth and growth rate did not vary significantly with litter size. Fleming and Rauscher interpret this to mean that these parameters are optimized by natural selection so that the most adaptive birth sizes and growth rates have evolved. Studies such as this are conducted on the assumption that the average litter size produces the maximum number of young surviving (Lack, 1946, 1947–48), however, there are other arguments concerning the evolution of litter (or clutch size) and these are discussed below.

Skutch (1949, 1967) argues that in stable environments animals (birds in particular) are close to a saturation density and they produce a litter (clutch) size which is adjusted to balance mortality and simply replace themselves – few young mean few visits to the nest and therefore fewer times when predators might be attracted. Fretwell (1969) argues that this apparently opposing argument to that of Lack and others may be reconciled

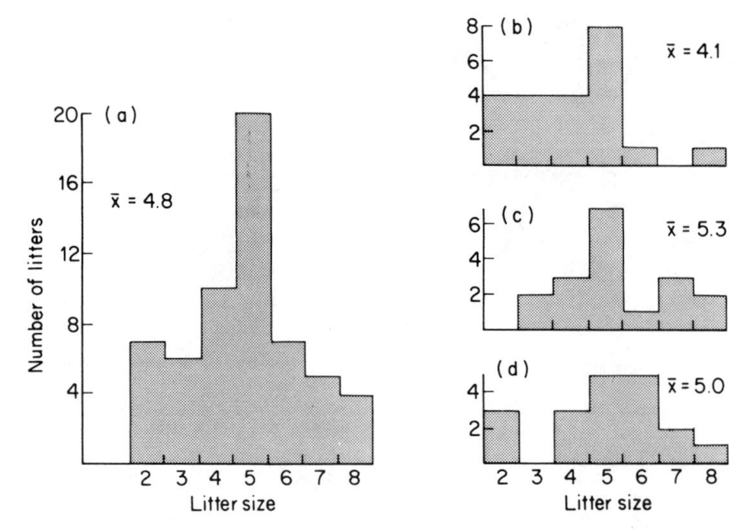

Fig. 4.4 Distribution of deer mouse litter sizes in the total samples **(a)** and in first **(b)**, second **(c)**, and third or later litters **(d)**. Redrawn from Fleming and Rauscher (1978).

if it is supposed that adults try to produce as many *breeding offspring* as they can, but that whenever mortality selectively impinges on excess offspring through social dominance or other effects, then the birth rate can be adjusted downward, not in order to balance the death rate but to maximize the number of surviving progeny. Arguments such as these highlight the role which bird studies have played in the development of theories of the evolution of litter size. There are so many other ideas but they are orientated particularly to birds (e.g. Klomp, 1970) and, as Stearns, (1976) points out, many of these arguments are now placed in a more comprehensive context. Different explanations may apply to different species, and factors such as behavioural complexity, parental care and interactions with other reproductive parameters will influence the evolution of the most productive litter size. These ideas are discussed later, but it is instructive to consider an argument which relates optimum clutch size to adult mortality that was put forward by Charnov and Krebs (1974). This assumes that there is a positive correlation between clutch size and adult mortality. For long-lived species large litter sizes reduce the parent's chances of surviving and reproducing in the future. Thus the lifetime contribution to future generations is optimized by producing a litter smaller than the most productive single litter size. This is described by the model shown in Fig. 4.5. It shows that if mortality of adults increases with clutch size then the optimum litter size is smaller than that predicted simply by juvenile mortality and litter size interactions. However, as Stearns (1976) points out, the position of the optimum clutch size depends upon the shape of the curve showing the trade-off between the reproduction and mortality of adults and this had never been measured!

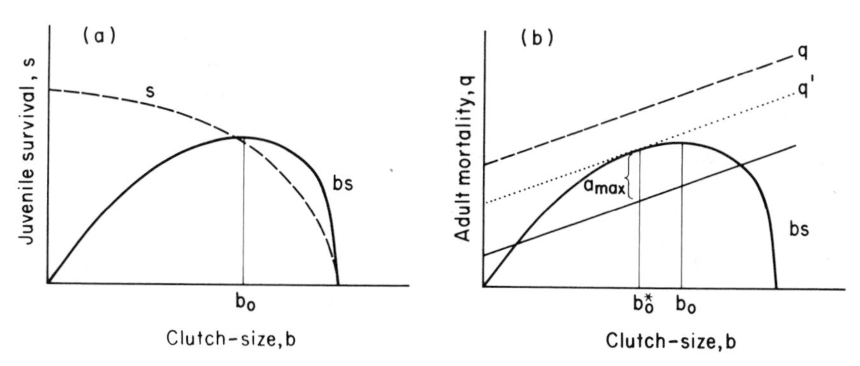

Fig. 4.5 The optimization of clutch size. The argument suggests that in **(a)** the relationship between clutch size, b, and survival rate for the first year of life can be shown by the broken line, s. The resulting production curve, bs, (solid line) has a single maximum at b_0. Under Lack's hypothesis this is the clutch favoured by natural selection. The impact of the cost of reproduction is illustrated in **(b)**. If adult mortality, q (dashed line), increases with b, then the clutch b_0^* that maximizes the measure of fitness, a, is always smaller than b_0. The optimal clutch is found by constructing the line parallel to q (q', dotted line) which just intersects the bs curve (solid line). In general, clutches smaller than b_0 will have a higher fitness than those larger. Redrawn from Stearns (1976) after Charnov and Krebs (1974).

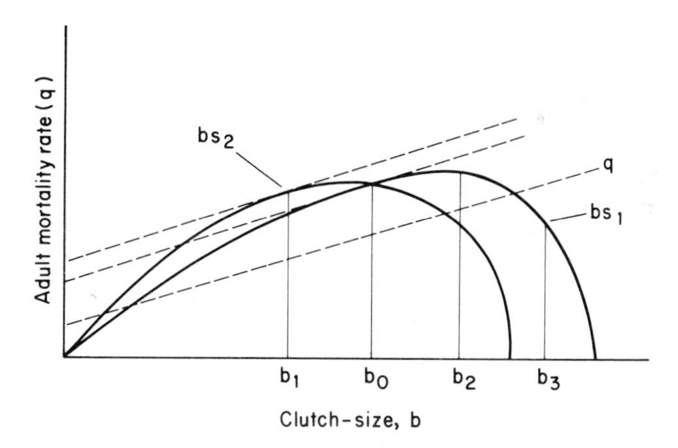

Fig. 4.6 The 'bet-hedging' hypothesis. Bet-hedging reduces average optimum clutch size. It is assumed in Fig. 4.5 that there is a trade-off between clutch size and the survival of young. If there is year-to-year variability in the optimal clutch size, selection favours bet-hedging so that a smaller clutch is an advantage. If b_0 is the optimal clutch size this year with production bs_1 and in the next year under different conditions production curve bs_2 operates, then a smaller clutch size, b_1, will offer maximum fitness, while a equivalently larger clutch, b_2, offers reduced fitness; a very large clutch, b_3, would result in a serious mortality risk to the parent. The analysis given depends on the shape of the bs and q curves, which are assumed to vary from species to species, place to place, and time to time. Redrawn from Stearns (1976).

The theoretical approach developed above forms part of an important theory proposed by Boer (1968) and Mountford (1973) which has become known as the 'bet-hedging' hypothesis. It simply refers to the hedging of bets in the face of uncertainty. If environmental conditions vary from year to year the animal cannot be certain at the time of reproduction what conditions will be like for the rest of the year. Thus a large litter may result in disaster with the litter and the adult dying but a smaller litter may result in some young surviving. This means that the optimal clutch size for one year may be variable between years because of variable juvenile mortality and so the lower litter size is selected – hedging the bet on the side of the smaller than optimal litter size (Fig. 4.6).

The above theories point the way towards a more sophisticated analysis of how litter size is determined than the simple food limitation hypothesis of Lack (1946, 1947–48). However, one must not lose sight of species which do show very simple relationships with environmental variables because of the severity of the climate. The South African rock hyrax (*Procavia capensis*) has a mean litter size which varies from 1.5 to 3.5, apparently with the rainfall and the season (Skinner *et al.*, 1977). Around Lake Menz in South Africa they found that hyraxes living in the wetter areas had a mean litter size of 3.5 but this was reduced to 2.4–2.6 in the poorly vegetated areas and drought conditions caused a further decline to 2.1 from a reduced ovulation rate and greater foetal wastage. It seems that the environment is so variable that the species may be incapable of adapting to it.

Semelparity and iteroparity

An important variable in the life-history of any organism is whether the individual reproduces once in its life (semelparity) or repeatedly during one or more breeding seasons (iteroparity). Semelparity is common among the bacteria, plants and invertebrates (Cole, 1954) but occurs only infrequently in vertebrates (lampreys, salmon and freshwater eels). The only mammalian examples of semelparity occur in the monooestrous marsupial mice of the genus *Antechinus* (Braithwaite and Lee, 1979), where all the males die after a two-week long mating period in early spring (Lee *et al.*, 1977) but some females survive after lactation to reproduce for a second time. In eutherian mammals it is only the degree of iteroparity which varies. The particular circumstances which have led to the evolution of semelparity are probably where the survival of adults to a second bout of reproduction is low, where the length of life is short (about a year or one breeding season) and of such a duration that the female will be able to raise one but not two litters (Braithwaite and Lee, 1979). Few small mammals survive to a second breeding season and possible candidates for semelparity in the eutherians are those species which live in habitats with only a short period for optimal reproduction and a short growing season. Species which come close to semelparity are a shrew (*Sorex vagrans*) and the rodents *Thomomys talpoides* and *Eutamias minimus* which produce a single litter at 3000 m in Colorado in comparison with 2 or more litters at lower altitudes. None of these species in the montane situation showed complete annual mortality but the survival of the shrew to a second breeding season was very low (Vaughan, 1969).

The advantages of iteroparity have been the subject of much theoretical debate since Cole's (1954) discussion of life-history phenomena. It has taken 20 years for other theorists to fully understand the initial arguments! Cole assumed that there was no difference between adult and juvenile mortality rates by assuming no mortality until after reproduction. He came to the conclusion that in an annual species the absolute gain in intrinsic population growth which could be achieved by changing to the perennial (iteroparous) reproductive habit would be exactly equivalent to adding one individual to the average litter size. This controversial result was criticized and countered by Murphy (1968) but not properly understood until Charnov and Schaffer (1973) looked at variation in both juvenile and adult mortality in relation to semelparous and iteroparous habits. They concluded on theoretical grounds that for an annual species the absolute gain in intrinsic population growth rate that can be achieved by a change to the perennial habit would be exactly equivalent to adding P/C individuals to the average litter size (where P = adult survival rate and C = survival of offspring in the first year). They show, as does Bell (1976), that Cole's conclusion is simply a special case where there is no pre-reproductive mortality (P/C = 1). In the more likely event of some mortality occurring before reproduction the advantage of iteroparity is equivalent to an increase in litter size (for a semelparous species) related to the ratio of adult survival to juvenile survival.

Thus these theoretical (mathematical) arguments reach the same conclusion as suggested by Braithwaite and Lee (1979). Semelparity evolves when adult survival is low in relation to juvenile survival, and iteroparity

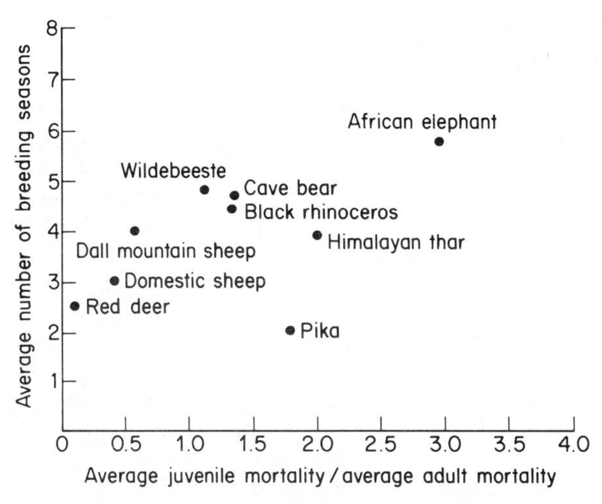

Fig. 4.7 The average number of breeding seasons for selected mammals is positively correlated with the juvenile/adult mortality ratio. Redrawn with modifications from Stearns (1976).

evolves under the opposite circumstances. Despite the absence of sexual and genetic aspects of the mathematical arguments these may not be necessary to understand the basic advantage of the evolution of iteroparity (Stearns, 1976). Stearns tests the general conclusion by analysing the relationship between average number of breeding seasons and a figure compiled from the average juvenile mortality divided by the average adult mortality (equivalent to 1/survival) of a number of species (Fig. 4.7). The data reproduced in the figure are only those from mammals and it is suggestive that the conclusions of Charnov and Schaffer are correct – however, it is not a proof. Many other environmental pressures may influence reproduction to select for iteroparity; among them may be the association of iteroparity with parental care when post-reproductive survival is favoured (Stearns, 1976) and the possibility that iteroparity may be selected for when the risk of total reproductive failure in any given year is significant (Holgate, 1967), when juvenile survival is variable (Schaffer, 1974) and also when maturity is delayed (Bell, 1976; Cole, 1954).

Co-adaptation of reproductive and life-history variables – *r*- and *K*-selection

The terms *K*-selection and *r*-selection were proposed by MacArthur and Wilson (1967) for two types of selection which were first described by Dobzhansky (1950). Dobzhansky argued that natural selection in the tropics operated in a different way from natural selection in temperate areas. In the temperate zones much of the mortality is relatively independent of the genotype of the organism and has little to do with the size of the population. However, in the relatively constant tropics most mortality is more directed, favouring individuals with better competitive abilities. Thus in the tropics lower fecundity and slower development help to increase competitive ability but in temperate zones selection favours high fecundity and rapid development. MacArthur and Wilson (1967) drew on these and later, similar ideas, to propose the terms '*r*-selection' for selection in environments favouring rapid population growth and '*K*-selection' for selection in saturated environments, favouring the ability to compete and avoid predation. *K* refers to carrying capacity (as in the logistic equation, see Chapter 5) and *r* refers to the intrinsic rate of natural increase. The characteristics which would be selected for and which correlate with each type of selection are shown in Table 4.2. The theory is qualitative not quantitative and allows comparisons only within limited taxonomic groupings (Stearns, 1976). It should not be treated as a mutually exclusive theory but as a *r*–*K* continuum so that an organism's position on the continuum might be visualized (Pianka, 1970). The *r*-endpoint represents the quantitative extreme (no density effects and no competition) and leads to high productivity, whereas the *K*-endpoint represents the qualitative extreme where density effects are maximal and the environment is saturated with the animal. Competition is keen, only a few extremely fit individuals are produced and this leads to the efficient utilization of resources. Figure 4.8 illustrates the relationships between the instantaneous rates of increase and the intensity of competi-

Table 4.2 Some of the correlates of *r*- and *K*-selection (from Stearns, 1967).

	r-selection	*K*-selection
Climate	Variable and/or unpredictable	Constant and/or predictable
Mortality	Density-independent; uncertain adult survival	Density-dependent; uncertain juvenile survival
Survivorship	Often Type III (Deevey, 1947)	Usually Types I and II (Deevey, 1947)
Population size	Variable in time, nonequilibrium; usually below carrying capacity; frequent recolonization necessary	Constant in time, equilibrium; at or near carrying capacity; no recolonization necessary
Competition	Often lax	Usually keen
Selection favours	1 Rapid development 2 High *r*-max 3 Early reproduction 4 High resource thresholds 5 Small body size 6 Semelparity 7 Increased birth rate	1 Slow development 2 Competitive ability 3 Delayed reproduction 4 Low resource thresholds 5 Large body size 6 Iteroparity 7 Decreased death rate
Length of life	Short, < 1 year	Long, > 1 year
Lead to	Productivity	Efficiency
Proportion of energy allocated to reproduction:	Relatively large	Relatively small
1 Mass of offspring/parent/brood;	Larger	Smaller
2 Mass of offspring/parent/lifetime;	Larger	Smaller
3 Size of offspring	Smaller	Larger
4 Parental care	Less	More

tion at the two ends of the continuum. The lower diagrams illustrate that at high population densities (high competition) the *K*-strategists have a higher actual instantaneous rate of increase and are therefore at an advantage but at low population density (low competition) the *r*-strategists are at an advantage.

It is important to realize, however, that the *r*–*K* continuum is not an 'all-embracing theory' and that high or low productivity may evolve for reasons other than *r*- or *K*-selection. In order to make predictions about the life-history strategy of a species it is important to consider as many determinants as possible; these may include the density of the population in relation to resources, the trophic and successional position of the population, the predictability of mortality patterns, and the predictability and

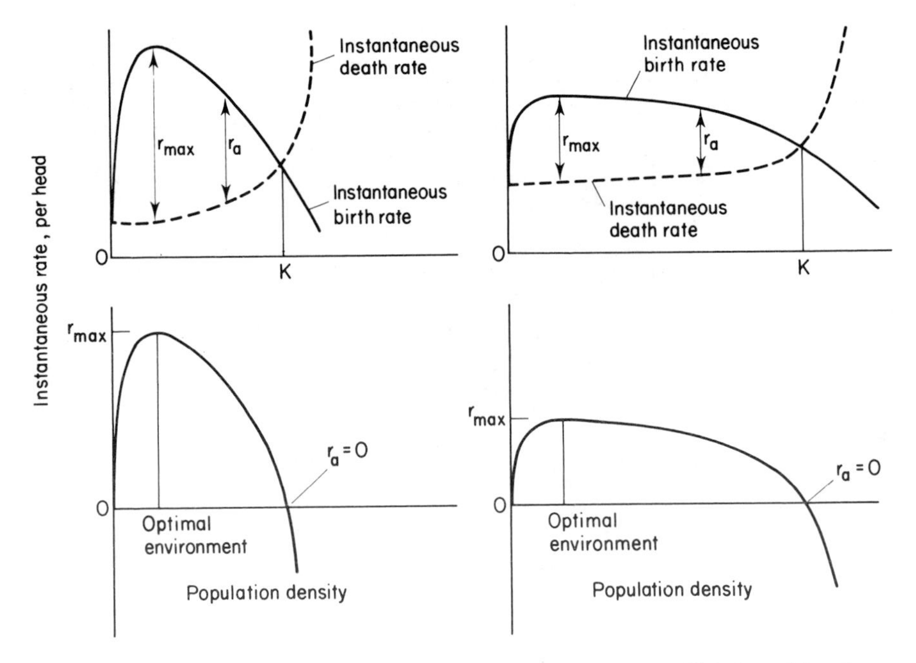

Fig. 4.8 Model relationships of various instantaneous rates of birth and death per head versus the intensity of competition (represented by population density) in an *r*-strategist (*left*) and in a *K*-strategist (*right*). At densities below the optimal environment (where rate of increase is maximal), the birth rate per head is low because probabilities of mating are reduced. The lower two curves represent the instantaneous rate of increase, r_a, being the difference between the instantaneous death rate and the instantaneous birth rate. Redrawn from Pianka (1972).

stability of the environment (Wilbur *et al.*, 1974; Fleming, 1979). The possible correlates of *r*- and *K*-selection and bet-hedging under different environmental and mortality conditions are summarized in Table 4.3. Unstable strategies such as those of the South African rock hyrax discussed earlier in this chapter are obviously not covered by these theoretical approaches!

An example of successful use of *r*- and *K*-selection theory is seen in a study of two species of voles, *Microtus pennsylvanicus* and *Microtus breweri* (Tamarin, 1978a). *M. breweri* was studied on an island off the coast of North America and compared with the nearby mainland species *M. pennyslvanicus*. On the island *M. breweri* are usually at high density and should therefore be relatively *K*-selected whereas on the mainland *M. pennsylvanicus* shows cyclic population fluctuations with severe declines at times and should thus be *r*-selected. In comparing the characteristics of the two species it was found that the island species showed a greater longevity, larger body size, faster development, smaller mean litter size, better parental care and less overt aggression; all except the faster development being '*K*-selected' attributes. Similar *K*-selected characteristics have been reported for the bank vole (*Clethrionomys glareolus*) on Skomer Island off

Table 4.3 The contrasting predictions of *r*- and *K*-selection and bet-hedging (from Stearns, 1976).

Stable environments	Fluctuating environments
r- and *K*-selection and bet-hedging with adult mortality variable	
Slow development and late maturity	Rapid development and early maturity
Iteroparity	Semelparity
Smaller reproductive effort	Larger reproductive effort
Fewer young	More young
Long life	Short life
Bet-hedging with juvenile mortality variable	
Early maturity	Late maturity
Iteroparity	Iteroparity
Larger reproductive effort	Smaller reproductive effort
Shorter life	Longer life
More young per brood	Fewer young per brood
Fewer broods	More broods

Wales where the voles are larger and have a smaller litter size and shorter breeding season than those on the mainland (Jewell, 1966). Not all island species conform to the *K*-selected predictions, however; for example deer mice (*Peromyscus maniculatus*) on the Gulf Islands of British Columbia show the expected larger body size of higher densities than on the mainland but also show rapid growth rates, high reproductive rates and poorer or similar survival to populations on the mainland (Sullivan, 1977), and are therefore relatively *r*-selected in their characteristics.

The *r*-, *K*-selection theory may be invoked to explain many taxonomic variations in reproductive and life-history characteristics. Fleming (1975) compared the annual probability of survival with the annual productivity of twenty species of tropical rodents (Heteromyidae, Echimyidae and Muridae), and the relationship between these two variables is shown in Fig. 4.9. The negative relationship between survival and productivity is interpreted as demonstrating that reproduction does entail a risk of mortality but at the same time it shows that there are some species of tropical rodent which are more '*r*-selected' than others, the low productivity and high survival species on the left of the scattergram contrasting with the high productivity, low survival species on the right of the scattergram. Comparisons such as this are taken further by French *et al.* (1975) in an effort to identify trends and patterns in the reproductive strategies of small mammals. They provide a similar comparison to that in Fig. 4.9, but for ecological (not necessarily taxonomic) groupings of small mammals. This is shown in Fig. 4.10 and compares survival rate with reproductive capacity (number of litters × litter size). There are three main groupings of species in the expected negative relationship between survival and reproductive

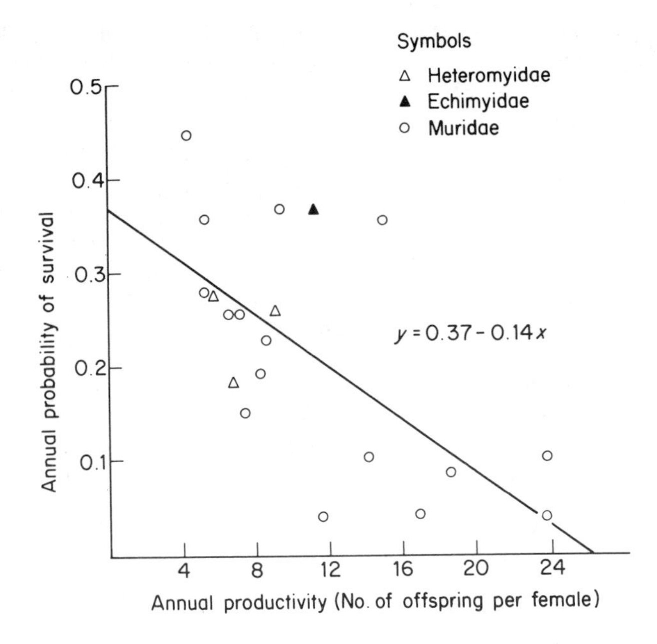

Fig. 4.9 Relationship between annual survival rate and annual productivity in 20 species of tropical rodents. Redrawn from Fleming (1975).

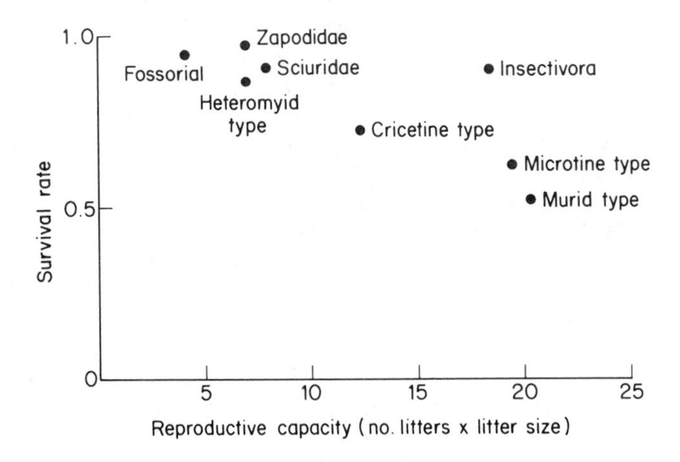

Fig. 4.10 Relationship between survival rate and reproductive capacity in 'ecological' groups of small mammals. Survival is the mean monthly rate for the group, and reproductive capacity is the mean seasonal production of young. Note that the groups of mammals share similar ecological characteristics and are not necessarily taxonomically related. Redrawn from French *et al.* (1975).

capacity (equivalent to productivity). At the '*K*-selected' end of the line are the fossorial rodents, the zapodids, the sciurids and the heteromyid types having low reproductive rates, high survival and usually showing low population densities. In the middle are the insectivores and the cricetine types usually showing moderate population densities, and at the '*r*-selected' end are the microtine types and the murid types which have poor survival, high reproductive rates and which often show high population density, although they may also show drastic declines. Whether selection acts in quite such a simple manner to bring together so many reproductive and life-history attributes as a result of one determining factor is open to question. However, in many cases the *r*- and *K*- attributes do appear to be co-adapted to particular circumstances and thus they have their place in evolutionary theory.

Post-reproductive individuals

Chapter 2 discusses the relationship between reproductive success and dominance. If dominance is related to age, and this is commonly so in mammals, then it might be expected that the older individuals would have greater reproductive success, particularly the males of the species. However, this is not always so. In the orang-utan, (*Pongo pygmaeus*) old males are very aggressive and eventually take no part in reproduction, in fact they are probably impotent (Mackinnon, 1974, 1977, 1978). Females avoid these old aggressive males although they do call loudly to attract females. Mackinnon interprets this behaviour as having a role in the protection and provision of the offspring of the old male by keeping other males away and guaranteeing the subadults sufficient space and food to grow up and breed successfully. The male's breeding strategy seems to be divided into two stages; as a subadult he fathers as many offspring as possible and as an adult he consolidates and protects his contribution to future generations. A similar strategy may even be seen in the female elephant, but how widespread it is is difficult to tell. Elephants apparently undergo a physiological change similar to the human menopause and females studied in the Murchison Falls Park South (Uganda) showed a 50% inactivity rate (not pregnant, oestrous cycling or lactating) in the 50–55 year age class and 100% inactivity in the oldest 55–60 age class. These barren females were often leaders of family groups and the groups presumably gain from the experience of the old female although she can no longer contribute directly to reproduction (Laws *et al.*, 1975).

5

The Description and Analysis of Population Dynamics

Population ecology relies on accurate observation and experimentation to provide the basic data for interpretation and analysis. The methods involved may be direct counts of the numbers of individuals or samples obtained from indirect methods such as transect counts, trappings, faecal pellet counts or other signs such as skeletal remains. Such methodology is described in Twigg (1975a, b, c), Flowerdew (1976) and Caughley (1977). Where the numbers of individuals in a population can be estimated or counted there is then the possibility of repeating the procedure and obtaining data on the life-history of the individuals, as well as quantifying the changes in population numbers from season to season. The study of survivorship or mortality in the population demands careful observation of live animals throughout their life span or analysis of the ages of dead samples. The time necessary to follow individuals throughout their lives makes this type of study difficult for some long-lived species and so short-cut methods, utilizing samples taken for age determination at one time, may be applied. This chapter considers some of the forms of analysis appropriate to mammalian populations and gives examples of their use.

Life table analysis

The population statistics used for life table analysis are simple time-schedules of births and deaths which give an average figure for mortality rates and birth rates for each age interval. These statistics allow changes in mortality or natality to be described by survivorship (the complement of mortality) and fecundity curves.

Life table data on survival (Table 5.1) show how many individuals of a cohort survive to the start of each age class; these data in the l_x column of the life table may show the figures as they are collected or may be more usually transformed to start with 100 or 1000 individuals. To collect the data for an 'age-specific' life table a cohort is followed from birth to death and the numbers (l_x) alive at the start of each unit period of time are noted or alternatively a note is made of the deaths during each period of time (d_x values in Table 5.1). In 'time-specific' life tables the population is sampled at one period of time and the asumption is made that the population is neither increasing nor decreasing, i.e. it has a zero growth rate and the population structure is stable. The decline in numbers from one age class to the next then represents the mortality over the time represented by each

Table 5.1 A life and fecundity table for the grey squirrel (*Sciurus carolinensis*) in North Carolina (data modified from Barkalow *et al.*, 1970). Note that l_x values refer to males and females jointly. © The Wildlife Society.

Age x (years)	l_x	d_x	q_x	e_x	m_x
0–1	1000	753	0.75	0.99	0.05
1–2	247	135	0.55	1.82	1.28
2–3	112	30	0.27	2.41	2.28
3–4	82	25	0.31	2.10	2.28
4–5	57	15	0.26	1.84	2.28
5–6	42	25	0.60	1.32	2.28
6–7	17	0	0	1.48	2.28
7–8	17	17	1.00	0.50	2.28

age interval. These figures can be used to construct the rest of the life table statistics and the survivorship curve but not, of course the rate of population growth which has been assumed to be 0. A common problem with this type of life table is that the young individuals may not be adequately sampled, especially if the dead remains of individuals have been collected to provide a d_x series; the skeletal remains of juveniles often disappear at a much faster rate than those of adults.

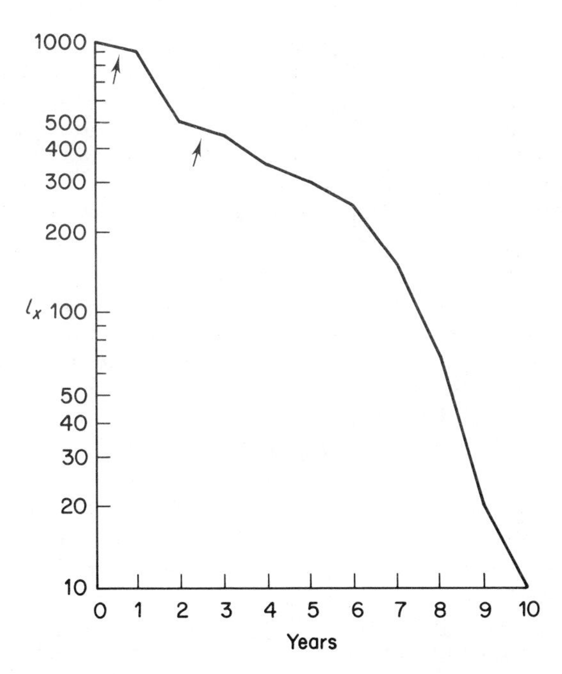

Fig. 5.1 Example of a logarithmic plot of a survivorship curve (l_x series) showing similar rates of mortality (q_x) from 0–1 year and 2–3 years of age (indicated with arrows).

Life table statistics refer to the numbers alive at birth and at the start of each age interval, and even if conversion factors (e.g. × 100/number alive at age 0) have been applied the parameters (q_x, e_x) derived from the l_x and d_x series will have the same value. The life table statistics starting from 1 at age 0 refer to probabilities so that the l_x series refers to the probability of living to the start of a given age interval, x, and d_x refers to the probability of dying during an age interval, x. The mortality rate, q_x, during each age interval, x, is simply d_x/l_x and the mean expectation of further life, e_x, is calculated from the formula:

$$e_x = \frac{1}{l_x} \sum_{x=y}^{n} \left(\frac{l_x + l_{x+1}}{2} \right)$$

where $y = 0$ or a higher age interval. Note that when survivorship curves (the change of l_x values with time) are plotted they are often represented on a logarithmic y-axis (Fig. 5.1) to show the rate of change of numbers of individuals between each age class. The change from 1000 to 900 gives a d_x value of 100 and a q_x value of 100/1000 = 0.1 and the change in l_x values from 500 to 450 gives a d_x value of 50 and a q_x value of 50/500 = 0.1; thus very different arithmetic changes in l_x values give similar q_x values and similar slopes between the changes in l_x values. In other words similar mortality rates will be indicated by similar slopes between the respective l_x values on a logarithmic scale because the subtraction of logarithms is equivalent to the division of the original numbers and 1000/900 gives the same result as 500/450; a straight line survivorship 'curve' will thus have a constant mortality rate throughout the species' life span.

The different ways of collecting l_x or d_x data from live or dead samples provide basic information for population studies and point to the times in an 'average' animal's life when mortality is particularly severe as well as providing a statistic, e_x, which gives the mean length of life for an individual in the population. The time-specific life table must be accepted as an approximation to the age-specific life table and checks on the accuracy of the assumptions involved in the time-specific table can be made by taking a further sample to assess population size for zero growth and similarity of population age structure. Examples of l_x and q_x curves from studies of mammals are shown in Fig. 5.2.

Fecundity schedules are often added to life table data; these are columns under the heading m_x, the average number of female offspring produced per unit time by a female in age class x; male reproduction is assumed to be similar to that of the female but unequal sex ratios and dissimilar survivorship between sexes should not be overlooked. To calculate the number of offspring produced by a population of *females* the l_x values (reduced to 1.0 at age 0) are multiplied by the appropriate m_x values to give an $l_x\, m_x$ curve for the population (Fig. 5.3). The area under this curve is approximated by $\Sigma l_x m_x = R_0$ = the net reproductive rate per generation. In many species of mammal the m_x values increase to an asymptote and remain at the peak level throughout the rest of their reproductive life, perhaps reducing slightly into old age. The parameter R_0 is equivalent to: number of daughters

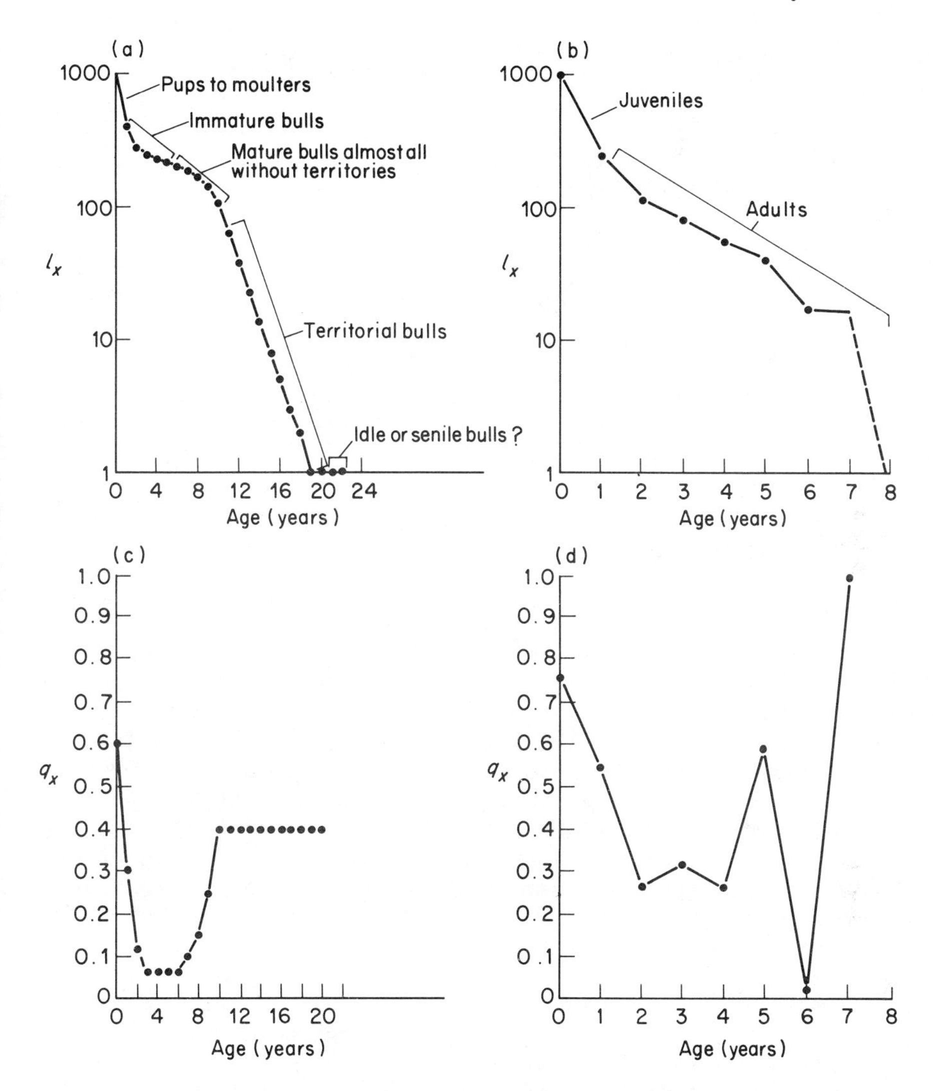

Fig. 5.2 **(a)** Survivorship curve for grey seal, *Halichoerus grypus*, bulls. (From data in Hewer, 1974.) **(b)** Survivorship curve for grey squirrels, *Sciurus carolinensis* (see Table 5.1). (From data in Barkalow *et al.*, 1970.) **(c)** Mortality rate curve for grey seal bulls. (From data in Hewer, 1974.) **(d)** Mortality rate curve for male and female grey squirrels (see Table 5.1). (From data in Barkalow *et al.*, 1970.)

born in generation $t + 1$/number of daughters born in generation t and if $R_0 > 1$ then the population is increasing and if $R_0 < 1$ it is decreasing. When $R_0 = 1$ the population is said to be stationary. Arising from these computations is T_c, the mean length of a generation (generation time) which is given by the following approximation:

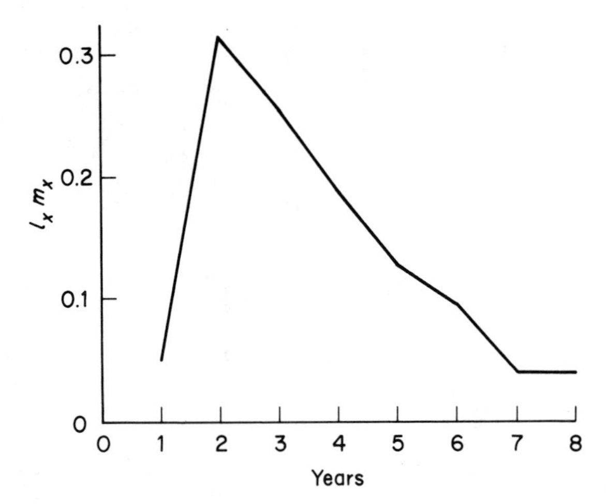

Fig. 5.3 Grey squirrel $l_x m_x$ curve calculated from average female survival and productivity values. Data from Barkalow *et al.* (1970).

$$T_c = \frac{\Sigma (x l_x m_x)}{\Sigma (l_x m_x)}$$

This statistic represents the mean period elapsing between the birth of the parents and the birth of the offspring, assuming a stable age distribution. It can be shown that a stable age distribution eventually will be achieved if a population is allowed to reproduce itself in a constant environment except when the population reproduces synchronously at a single age. When the stable age distribution is achieved the proportion of individuals in the different age classes remains the same as long as the l_x and m_x values are not changed whether or not there is an increase or decrease in numbers.

Rates of increase

Life table data or similarly derived figures from population studies may be used to calculate the rate of growth of the population. However, before the details of the calculation are considered, it is necessary to understand the theory of exponential growth and its use in population ecology.

Exponential, logarithmic or geometric growth is simply the mathematical way of expressing growth by the law of compound interest. Interest in money terms is added to the capital each year and the next year interest is calculated on the new total. Growth such as this is described by the exponential curve (Fig. 5.4) $y = e^x$, where

$$e^x = 1 + x + \frac{x^2}{(2 \times 1)} + \frac{x^3}{(3 \times 2 \times 1)} \ldots\ldots \text{etc.}$$

Simply pressing the 'e' button on a calculator shows that $e^0 = 1$, $e^1 = 2.71828$, $e^2 = 7.38906$, $e^3 = 20.08554$, and so on, so that the curve of $y = e^x$

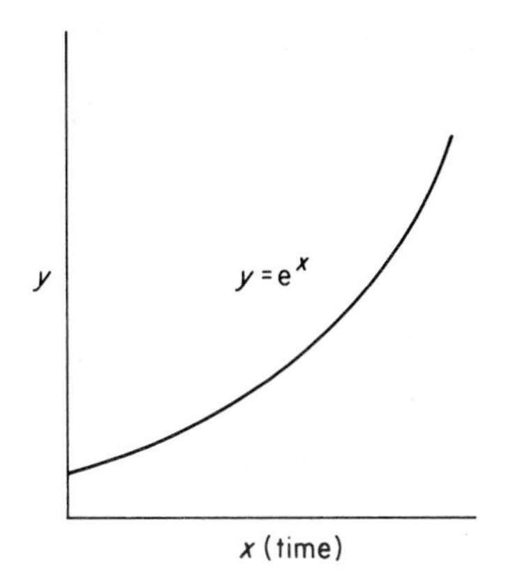

Fig. 5.4 The exponential growth curve, $y = e^x$, also described as $N = e^t$ or $N_t = N_0e^{rt}$ where population number, N, increases with time, t, in an exponential fashion and r is the instantaneous rate of increase.

will track a sharply increasing path with increasing values of x, the time axis. As the curve is continually increasing the rate of growth is continually changing and so it is necessary to calculate an instantaneous rate of growth by means of calculus.

If we use an ecological notation N, for the number in the population then differential calculus uses the notation DN to denote a small increment in N and this takes place during, or in other words is accompanied by, a small increment in time, t, denoted by Dt (see Fig. 5.5). Thus between time 0 and time t:

$$\frac{\mathrm{DN}}{\mathrm{Dt}} = \frac{(N_t - N_0)}{t - t_0}$$

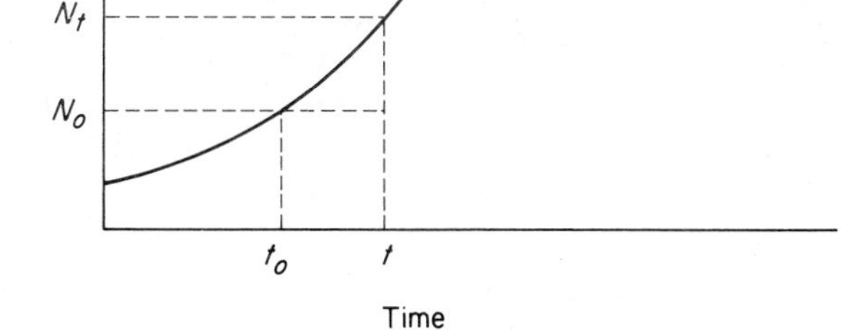

Fig. 5.5 Exponential curve showing the calculation of the rate of change of N between time t_0 and t.

The value DN/Dt will be different at time t_1 and t_2 and so the formula above is not generally applicable for the description of the rate of change in numbers over similar time periods unless the line is straight. However, in calculus, the rate of change of N at one instant of time is defined as dN/dt and as DN/Dt gets smaller and smaller it will approach dN/dt so that in the limit, when DN and Dt approach zero:

$$\frac{\mathrm{DN}}{\mathrm{Dt}} = \frac{dN}{dt} = \text{instantaneous rate of change of } N \text{ at time } t.$$

If the instantaneous rate of change is defined as 'r' (often called the intrinsic rate of natural increase) then

$$\frac{dN}{dt} = rN \quad \text{or} \quad r = \frac{1}{N} \times \frac{dN}{dt}$$

r represents the difference between the instantaneous birth rate b and the instantaneous death rate d per individual or $r = b - d$. The differential equation for r can be integrated (see Pielou, 1974) to give the following:

$$N_t = N_0\, e^{rt}$$

where N_0 = numbers at time 0 and N_t = numbers at time t. This expression can be used to obtain values of r over a single time interval and is easily evaluated with a pocket calculator.

If breeding shows a single birth pulse during each time period or is seasonal with only one litter each season then a simpler equation can be derived to describe the increase in the population which is called a 'finite' increase rather than an instantaneous increase. Thus if B = the finite rate of birth and D = the finite rate of death per individual per unit time then:

$$N_{t+1} = N_t + (B - D)N_t$$
$$N_{t+1} = N_t(1 + B - D)$$
$$N_{t+1} = \lambda N_t, \text{ where } \lambda = 1 + B - D$$

λ is defined as the finite rate of increase or the factor by which the population increases during a single time interval. By changing the time period in the previous equation so that

$$N_{t+1} = N_t\, e^r$$

the two equations for the instantaneous rate of increase and the finite rate of increase can be combined to give the following identity:

$$\lambda = e^r \quad \text{or} \quad r = \log_e \lambda \quad \text{(note } \log_e \text{ is the log to the base e)}$$

If the unit time interval is chosen so that $r = \log_e \lambda$ then a seasonally breeding population would increase as though growth occurred continuously during the period. The last equation giving r in terms of λ is often used to work out the value of r given λ or the change in numbers over a period of time. Finite growth rates vary in a different way to instantaneous growth rates as shown in Table 5.2.

The intrinsic rate of natural increase, r, is a characteristic of a species

Table 5.2 Instantaneous rates and finite rates of increase (see text for further derivations). These are nearly complementary when rates are very small. The table shows how they diverge as the rates become large and illustrates the change in size of a hypothetical population which starts at 100 organisms and increases or decreases at the specified rate for one time period (from Krebs, 1985). © 1985 by Harper & Row, Publishers Inc. Reprinted by permission of Harper & Row, Publishers Inc.

Percent change	Finite rate	Instantaneous rate	Hypothetical population at end of one time period
-75	0.25	-1.386	25
-50	0.50	-0.693	50
-25	0.75	-0.287	75
-10	0.90	-0.105	90
-5	0.95	-0.051	95
0	1.00	0.00	100
5	1.05	0.049	105
10	1.10	0.095	110
25	1.25	0.223	125
50	1.50	0.405	150
75	1.75	0.560	175
100	2.00	0.693	200
200	3.00	1.099	300
400	5.00	1.609	500
900	10.00	2.303	1000

when it can be measured in a constant environment, but it will obviously change with changes in environmental conditions. It is relatively easy to find the value of r under controlled conditions and this has been done for the field vole (*Microtus agrestis*) by Leslie and Ranson (1940); in the laboratory the value of r was 0.0877 per week and this is often denoted as r_{max}, indicating that it is the maximum value which may be attained. Using the equation $r = \log_e \lambda$ is only an approximate way of finding r, a better way is to solve the following equation, substituting trial values of r until the appropriate value is found.

$$\sum_{x=0}^{\infty} l_x m_x e^{-rx} = 1$$

Sigmoid growth curve

The sigmoid or logistic growth curve is a simple model of population growth with an upper limit to the increase (Fig. 5.6). The model assumes that growth occurs in a limited environment so that a 'carrying capacity' is reached where no further increase can take place. It is commonly observed that the rate of population increase is seldom exponential for very long, and one factor which may prevent further growth is the number of individuals already present. Thus as density rises the presence of other organisms reduces fertility and longevity, and further increase is limited by the amount of food or other resources available to the population. The sigmoid

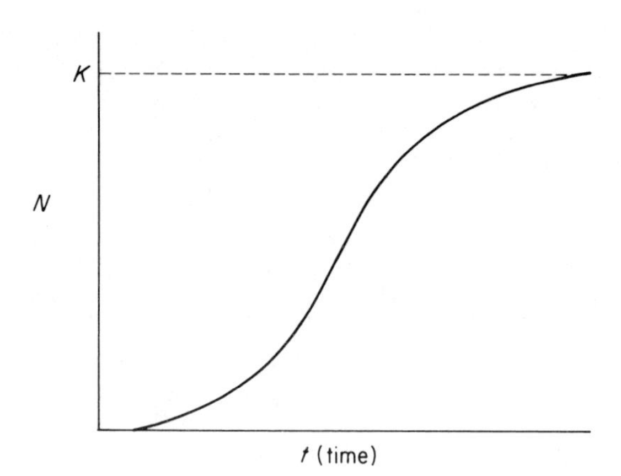

Fig. 5.6 Logistic growth curve which reaches carrying capacity K, when the instantaneous rate of growth $r = 0$.

growth curve represents the type of population increase which does not exceed a maximum level and which approaches this level smoothly. Strictly speaking it represents the type of increase and population asymptote that can be sustained by a limiting but renewable food resource and it is clear that this will not apply to many mammal populations, which by eating food today affect their food supply tomorrow (Caughley, 1976). However, the model is commonly used to represent growth to a maximum level and it is defined as follows:

$$\frac{dN}{dt} = \frac{rN(K = N)}{K}$$

where N = population number, t = time, r = intrinsic rate of natural increase and K = the upper asymptote or maximum level of N. Thus the rate of increase of the population per unit time increases and then decreases with increasing population size and is given by $r \times$ the population size, N, as before but this is also multiplied by the unused opportunity for further population growth $(K - N)/K$. This equation can also be integrated to give:

$$N_t = \frac{K}{(1 + e^{a - rt})}$$

where a is a constant defining the position of the curve in relation to the origin of the graph axes. Further models of population growth are discussed in Chapter 6 and described in many texts such as May (1981) and Elseth and Baumgartner (1981).

Key-factor analysis

This form of analysis, developed by Varley and Gradwell (1960), has been used predominantly for invertebrate populations and despite its critics (e.g. St Amant, 1970; Murdoch, 1970; Putman and Wratten, 1984) it does have its uses in helping to understand the population dynamics of vertebrates such as birds and large mammals (Podoler and Rogers, 1975; Hill, 1984; Clutton-Brock *et al.*, 1985).

The analysis is best applied to a population with discrete, non-overlapping generations where reductions in numbers by various mortalities occur sequentially and where these are often linked to developmental stages or to particular times of the year. A number of years of study are required so that the variation in the mortalities is fully described. The results of the analysis help to distinguish which mortality (the so-called key-factor) is responsible for the major fluctuations in population numbers and which, if any, mortalities exert a density-dependent (regulating) influence on population numbers and so help in maintaining numbers about an equilibrium level.

The population data can be expressed so that a hypothetical maximum number of young or the actual number of young produced each year is presented at the start of each annual life table. Then follow reductions in numbers, the first two (if calculated) being a hypothetical loss from failure to breed and a failure to reach the maximum clutch or litter size, then an actual loss during the juvenile life stage, possibly a loss between weaning and taking up a territory, an overwinter loss, and so on, according to the biology of the species, ending up with a number surviving to breed in the following year (if the biology of the species allows such breeding) from which, with the previous years' surviving adults, the following year's reproductive potential is calculated.

These reductions, initially linked to failures to breed (if calculated) and then to sequential mortalities, are transformed into logarithms (base 10) and the loss at each stage or 'killing power' of each mortality is defined as $k_1, k_2, k_3 \ldots k_n$ where $k_1 = \log_{10}(N_0/N_1)$, $k_2 = \log_{10}(N_1/N_2) \ldots k_n = \log_{10}(N_{(n-1)}/N_n)$. These equations simplify to:

$$k_1 = \log_{10}N_0 - \log_{10}N_1,\ k_2 = \log_{10}N_1 - \log_{10}N_2 \ldots k_n = \log_{10} N_{(n-1)} - \log_{10} N_n$$

The total mortality through each annual series of population statistics (the *generation mortality*) is the sum of the individual k values:

$$K = k_1 + k_2 + k_3 \ldots + k_n$$

This life table, its logarithmic transformations and k values adding up to the generation mortality (K) are illustrated in Table 5.3. Gains through immigration between two stages may be expressed in k values as negative reductions (increases), but will only be shown in special cases of the analysis where the population is assumed to be 'open', allowing immigration and emigration.

The k values and K for each year of study (preferably five or more years) are plotted against the year to allow the comparison of the annual variation in each k value with the annual variation in the generation mortality (K) (see Fig. 7.16). The k value which is on average the largest and which most closely parallels the generation mortality is called the 'key-factor' and is initially identified by eye. This is the mortality which usually causes the

Table 5.3 The calculation of k values for sequential losses of a hypothetical population during one year.

Stage	Loss from	No.	Log_{10} no.	k-value	Arithmetic reduction
Potential births		1000	3.00		
	k_1 Failure to breed			0.10	× 0.80
Newborn young		800	2.90		
	k_2 Failure to wean			0.20	× 0.63
Weaned young		500	2.70		
	k_3 Failure to survive winter			0.30	× 0.50
Overwintered young		250	2.40		
	k_4 Failure to gain a breeding territory			0.10	× 0.80
Breeding adults		200	2.30		
		Generation mortality = K =		0.70	

greatest change in numbers from one year to the next and accounts for most of the change in the amount and direction of K. For greater precision, it is possible to check this eye correlation by regression techniques (Podoler and Rogers, 1975) as sometimes the key-factor changes with time or two mortalities may be very similar in their action. There may not even be a key-factor!

The individual sets of k values can also be tested to see if a particular mortality is likely to be acting in a density-dependent manner and so be helping to 'regulate' the population about an equilibrium level. This is done by plotting each set of k values against the number on which they acted (usually the logarithm of the number, to help linearity). If the slope of the regression line is significantly greater than 0 then density-dependence may be suspected. Fig. 5.7 shows how the results may be interpreted as under-compensating density-dependence, when the mortality is not strong enough to bring numbers fully back to equilibrium, as perfectly density-dependent (slope of $b = 1$) and as over-compensating with a positive slope greater than 1, when the mortality will tend to 'overshoot' the equilibrium level. Even delayed density-dependence (Fig. 5.8) may be revealed from a time plot of the k value so that higher mortalities are expected in the year after the higher densities and a circular track of k is shown.

These relationships between mortalities and population numbers have a general relevance outide key-factor analysis in the search for density-dependence. However, the test should always be treated with caution because the two variables are often not independent of each other and further tests are necessary to support a suggestion of density-dependence (Varley *et al.*, 1973).

Unfortunately, this form of analysis does not tell the researcher the cause of the mortality. Ideally, the reason for the mortality will be known or else a measure of the cause of the mortality will be positively correlated with the k value; this can be done in the case of, for instance, rainfall or predation.

Further inferences can be drawn from tests for density-dependence. In

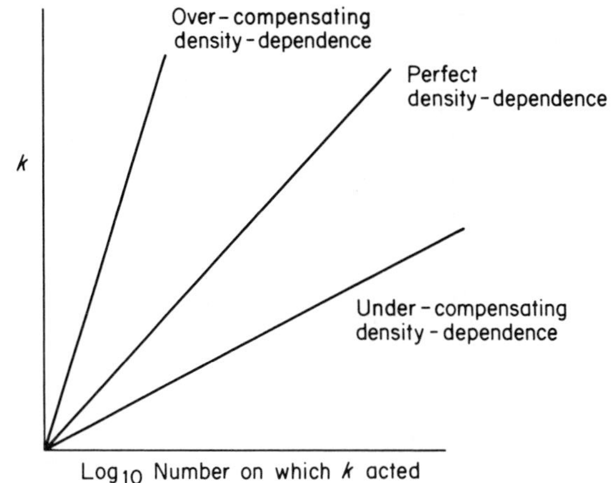

Fig. 5.7 Plots of k values against Log_{10} Number on which the k values acted, showing degrees of density-dependence in the relationship.

simple terms, a line with a slope of 1 in Fig. 5.7 could be said to indicate 'contest competition' (Nicholson, 1954) where each successful animal gets all it requires and the unsuccessful animals get insufficient for survival and reproduction: the slope of the line representing the variation of k with the numbers on which it acted starts at the point where resources begin to be restricted, such as a limited number of breeding sites, territories, and so on. In 'scramble competition' there is a slope greater than 1 and some or possibly all of the resource is taken by the competing animals but it is not enough to sustain the whole population. Thus there may not be enough resource for the survival of each individual and many may die. If a resource such as food is shared in this way and it is insufficient for maintenance then heavy mortality will ensue; however, there may be intermediate situations where a few individuals obtain enough food and so the slope of the plot would not be as steep.

In mammal populations with highly developed social organizations it seems unlikely that scramble competition would be found. However in species such as the red deer (*Cervus elaphus*) this seems to be the case (Clutton-Brock *et al.*, 1985). Also, in rabbit (*Oryctolagus cuniculus*) populations living in sand dunes scramble competition of males for both food and females is suggested (Cowan and Garson, 1985).

To complete the section on key-factor analysis it is worth considering the various objections to its use. Putman and Wratten (1984) give a comprehensive list of the pit-falls in key-factor analysis and those relevant to vertebrate populations are given below.

(1) The likelihood of mortalities overlapping in their action on the population should be considered; thus a disease may be acting on an individual at the same time as predation and the cause of death, although put down to predation, may be mainly the result of the disease. Does disease affect the

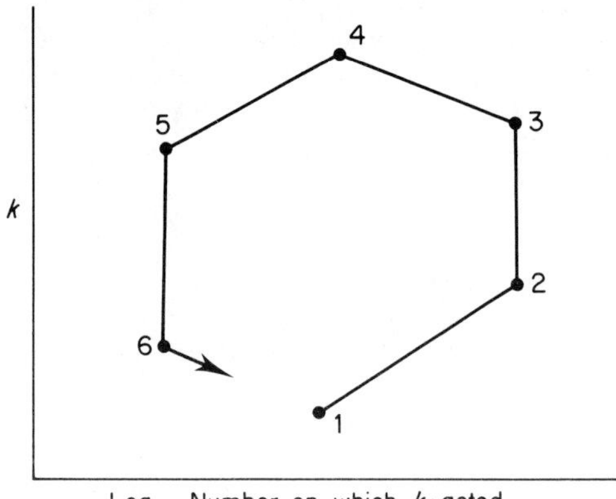

Fig. 5.8 Plot of k values against Log_{10} Number on which k values acted for a sequence of years, 1–6, showing a delayed density-dependent relationship.

probability of predation? Indeed, if disease is acting on the population, is mortality from this cause going to be detected?

(2) The ordering of the mortalities may be incorrect and arbitrary and some mortalities may be so 'wide' that they contain and hide many others; thus 'winter disappearance' may tell us little of the real losses to the population during a long period of time.

(3) If new sources of loss are recognized during the course of a population study it is difficult to incorporate these into the analysis of the previous population data, which was obtained without this extra information, and so further years of study may be needed.

(4) The relationship between the variables in testing for density-dependence, for instance, may be non-linear, perhaps starting only at a threshold level of density. This type of relationship is illustrated by Southern's (1970) analysis of tawny owl (*Strix aluco*) breeding losses where there is a reduction in losses as rodent density increases, but after rodents reach about 100 per 4.86 ha the losses remain at a minimal level (Fig. 5.9). It is also possible that the relationships plotted could be curvilinear as competition effects sometimes are.

(5) The population processes may change during the study due to changes in the environment, such as interactions with other animal populations, or the mortalities may act in different ways at different population densities. Simulation models may give more reliable conclusions and examples of this type of analysis are found in Chapter 8, p. 185.

Understanding the effects of environmental variables

In the study of vertebrate populations the environment may be an important influence affecting numbers, either in a positive or negative manner. If the

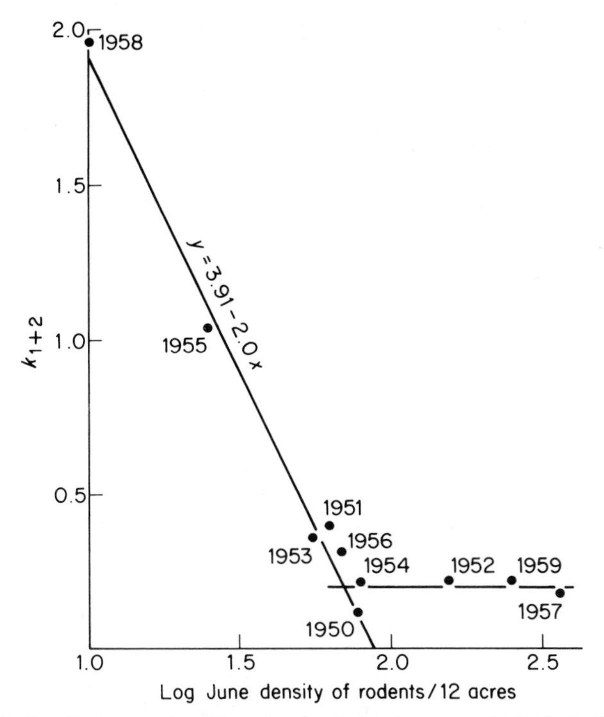

Fig. 5.9 Relation between density of rodents on 12 acres (4.86 ha) of woodland in June and tawny owl 'losses' through failure to breed and to achieve maximum size of clutch (k_{1+2}). After Southern (1970). By permission of the Zoological Society of London.

factors involved can be measured or manipulated then it may be possible to get closer to understanding how population numbers are regulated or limited, or simply modified by these factors.

Cause and effect

Ideally a change in mortality or population statistic should be attributable to a cause. This may be assessed in a number of ways in scientific work and will have variable accuracy according to the methodology (Romesburg, 1981). Simply finding that one environmental variable is positively correlated with a mortality or other population statistic, can, by the laws of induction, be used to associate the two in a causal fashion. The more this correlation is observed, the more credible the association suggested. If only past data are available the method of retroduction is applied where hypotheses about processes may explain the observed facts; however, other hypotheses may explain the facts equally well. These arguments are often supported by statistical correlation analysis but the confounding of the correlation of x and y because a third factor may be affecting both x and y in the same way is often overlooked. To be able to say that a change in variable x

causes a change in variable *y* will require an experiment showing changes in *y* following changes in *x*. In addition, if in the control situation a change in *x* was not made and a change in *y* did not occur then there is definite evidence that changes in *x* cause changes in *y* and repetition of this cause and effect relationship leads the experimenter to accept the truth of the relationship. This is much better scientific evidence than inductive or retroductive associations because the experiment allows the hypothesis to be properly tested.

Manipulation of populations

If experimental manipulations of mammalian populations are to be carried out to test an hypothesis then difficulties often arise if this is to be done in the field. This may be because of lack of manpower preventing replication and often this will limit the experimenter to one experimental and one control situation – being able to look at interactions between experimental treatments or being able to replicate experiments are almost unheard of in field experiments on mammals! The basic problems are unlikely to change in the future but experimenters should be encouraged to make the best of the situation by considering the reversal of control and experimental areas to eliminate site biases and introducing some measure of replication if at all possible.

6

The Population Dynamics of Mammals

Commonness and rarity

Mammalian population ecology is built around the study of fluctuations in population numbers together with investigations into population structure and demography, feeding, breeding, movements and social organization; habitat preference and interactions with other species also come under scrutiny. All these aspects of mammalian ecology may affect the fluctuations in numbers. The answer to why a species should be common or rare may be found under any one of these headings. In addition there may be fluctuations which are so marked that the species in one locality is common at one time and rare at another; however, the classification of species into common or rare usually refers to the long-term density to be found in a particular area, disregarding short-term fluctuations in numbers.

It is clear that habitat restriction will determine whether a species is common or rare in a particular area and that wherever mammals are studied there are species which are dominant in the species list and others which are 'occasional' or rare. Thus in rough grassland in Britain the field vole (*Microtus agrestis*) is likely to be at the top of the mammal species abundance list and the water shrew (*Neomys fodiens*) will probably be at the bottom. Similarly, in fenland sedge fields the bank vole (*Clethrionomys glareolus*) will be at the top and the water shrew will again be at the bottom (Flowerdew *et al.*, 1977). Species lists from any habitat will have similar common and rare species: it is easy to understand the social organization of a common species, but is a rare one really permanently living at a low density and functioning (breeding) as a population, or are such species really representatives from populations at a higher density elsewhere which are dispersing?

Even within one habitat a species may be very abundant for a time and then decline in density. This appears to have happened to the yellow-necked mouse (*Apodemus flavicollis*) in a Gloucestershire woodland (Montgomery, 1985) (Fig. 6.1). During the early 1970s the yellow-necked mouse was (unusually for British woodland) the most abundant species of rodent present, outnumbering the wood mouse (*Apodemus sylvaticus*). However, during the late 1970s and during the early 1980s the situation became more 'normal' with a reversal of the abundance of the two species. It is suggested that a change in habitat associated with the loss of many elm trees (*Ulmus* spp.) and a change in the abundance of other tree species present has influenced the abundance of the yellow-necked mice. In other British

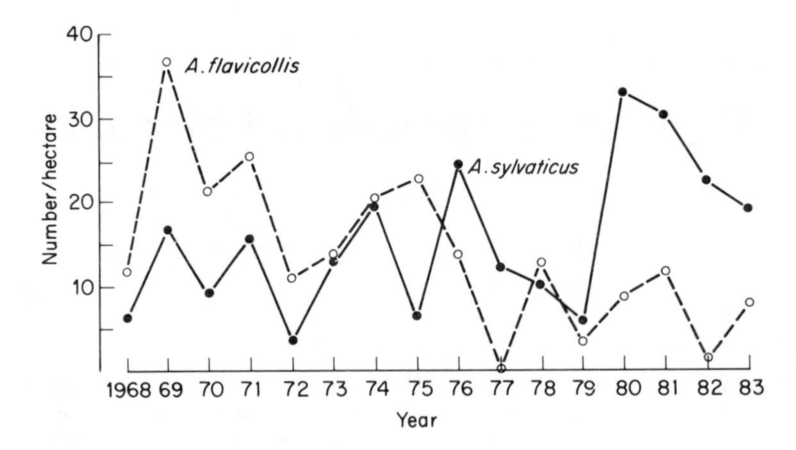

Fig. 6.1 Density of wood mice, *Apodemus sylvaticus*, and yellow-necked mice, *A. flavicollis* on grid M in Woodchester Park, Gloucestershire, 1968 to 1983. Numbers were estimated in June. Redrawn from Montgomery (1985).

woodlands in the south and south-west it is usual to find yellow-necked mice at very low densities in comparison with wood mice whereas on the continent of Europe such as in Poland it is more usual to find the yellow-necked mouse as the most common species with other *Apodemus* species less common.

Another example of a species with long-term changes in population density is the rabbit (*Oryctolagus cuniculus*) and various other species which were directly or indirectly affected after the introduction of the myxomatosis virus into Britain in 1953 (for review see Sumption and Flowerdew, 1985). On an estate in Hampshire, Tapper (1982) collected the figures for the numbers of weasels (*Mustela nivalis*) and stoats (*Mustela erminea*), killed by three gamekeepers from the start of the myxomatosis epidemic. Stoats initially declined, probably because of the lack of prey, but recovered in the 1960s as they adapted to the new situation. However, the weasels showed a remarkable increase in numbers until the early 1970s and then a decline. During this period the numbers of rabbits shot on the estate had increased by an average of 14% per annum and this obviously contributed to the recovery of the stoat population. It is suggested that after the decline in the rabbit population due to myxomatosis much more long-grass habitat was available for field voles and that this allowed the weasel population to increase. After the increase of the rabbit population during the 1960s the available habitat for field voles decreased as grassy banks were grazed bare by the rabbits and this could explain the decrease in the number of weasels being killed during the 1970s.

Interactions with other species may be an important determinant of which species are common and which are rare in a particular habitat. In Britain the grey squirrel (*Sciurus carolinensis*) has replaced the red squirrel (*Sciurus vulgaris*) in most of the deciduous woodland and even in parts of its coniferous woodland habitat (K. MacKinnon, 1978). It has been difficult to pinpoint the nature of the interaction between the two species, but it seems that wherever the grey squirrel is able to colonize it has been very difficult for the red squirrel to re-establish itself; thus a common species is becoming a rare one and vice versa.

Species living at different trophic levels will have densities which vary according to the body weight of the species and the available food density, among other factors. Thus it would be expected that a predator species would occur at a much lower density than many of its prey species; carnivores usually occur at a lower density than herbivores. A population ecologist will not expect a much greater return in live-trapping a carnivore than one catch per 100 trap-nights whereas the small rodent catch might be expected to be in excess of ten per 100 trap-nights. This statement is obviously a gross oversimplification, but serves to illustrate the difference between trophic levels in population density.

Stability and instability

Populations of organisms are not always in optimal habitats where fluctuations may be relatively damped. Thus at the edge of a species' range the population is occupying very little suitable habitat and the fluctuations will be more violent, perhaps with occasional outbreaks and extinctions.

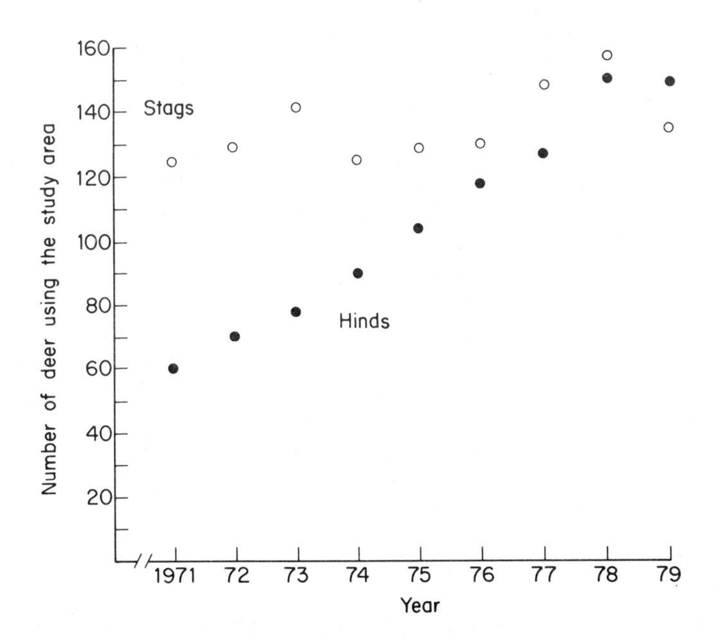

Fig. 6.2 Total number of red deer hinds and stags aged one year or more that regularly used a study area on the Island of Rhum (Scotland) from 1971 to 1979. Redrawn from Clutton-Brock *et al.* (1982).

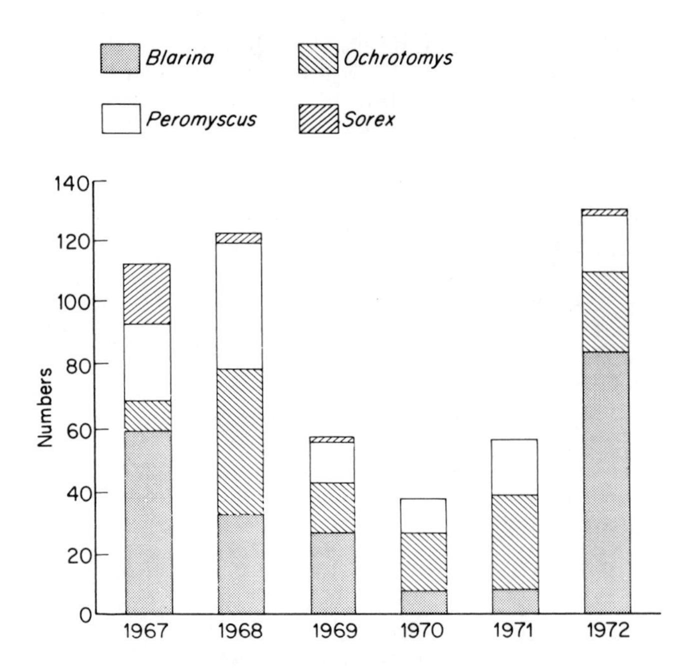

Fig. 6.3 The number of short-tailed shrews (*Blarina brevicauda*), golden mice (*Ochrotomys nuttalli*), cotton mice (*Peromyscus gossypinus*) and Southeastern shrews (*Sorex longirostris*) caught on a standard grid during 18 consecutive days of removal trapping during the summers of 1967–1972. Redrawn from Smith *et al.* (1974).

Similarly it is conceivable that a population within a favourable habitat will fluctuate less than those in a less suitable habitat. Thus species may be found which are stable in one area but unstable in another (see Chapter 7). In addition, the inherent differences between species should not be overlooked; K-selected species (Chapter 4) will show relatively slow changes in population density in comparison with r-selected species. Red deer (*Cervus elaphus*) on the island of Rhum off the coast of Scotland have slowly increased in numbers since management (culling) stopped in 1972–73 from 19.8 deer/km^2 in 1971 to 30.5 deer/km^2 in 1980 (Clutton-Brock *et al.*, 1982); the differential increase in stags and hinds is shown in Fig. 6.2). In contrast, populations of small mammals in eastern USA fluctuate markedly from one year to the next (Smith *et al.*, 1974) (Fig. 6.3). Long-term declines in populations occur often as a result of man's intervention by hunting or by habitat destruction; the decline in the otter (*Lutra lutra*) population in England (Chanin and Jefferies, 1978) is probably a result of one or both of these factors and perhaps others such as disturbance or pollution (Fig. 6.4).

Fluctuations in mammalian populations may be irregular, annual or multiannual. In the regular cyclic (multiannual) fluctuations found in many rodents, lagomorphs and their predators in northern latitudes (see Chapter 7) rodents tend to cycle at a period of 3–4 years and lagomorphs at about 10

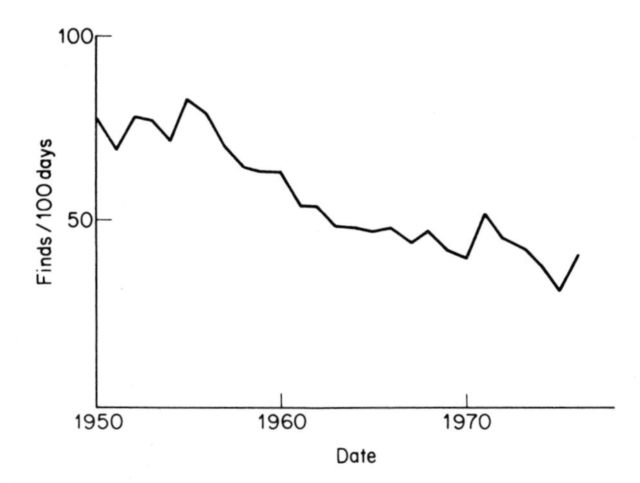

Fig. 6.4 Hunting success (finds per 100 hunting days) of all British Otter Hunts for 1950–1976. Redrawn after Chanin and Jeffries (1978).

years. Thus mammalian populations present a range of characteristics in their dynamics which ecologists hope to explain. There are likely to be different explanations for different species or perhaps for the same species in different parts of its range, as has been found for a number of insects. The facts that many populations do remain relatively stable within broad limits, that they do not increase for long periods, if ever, at the rate of increase suggested by the theoretical value of r under favourable conditions, and that they do not often overeat their food supply, all suggest that some density-dependent influences act upon population numbers to stabilize them. How common this is can be seen in the later sections of this chapter and the next.

The case for density-dependence

The arguments concerning the importance of density-dependence in animal population ecology are now of historical interest rather than the burning issues which they were in the '50s and '60s. Ecologists often suggest that the simple explanations of events are likely to be the correct ones, but this has also led to thoughts of single explanations for many population phenomena observed in many different species, and this is the root of many of the arguments of the past. Today the arguments are regarded as simply the result of individuals taking a dogmatic view of issues, but despite this a brief consideration of the different view points will, hopefully, help the reader to understand the problems involved in population ecology.

The arguments over population ecology have been fully discussed in many text-books and have a different emphasis when invertebrates are considered (see Itô, 1978; cf., Lack, 1966 for vertebrates). Nicholson (1933) is commonly quoted as being the proponent of facultative (density-dependent) factors stabilizing (i.e. regulation, see Chapter 5) population numbers. In fact both Nicholson (1933) and Smith (1935) followed Howard and Fiske (1911) in emphasizing the importance of density-dependent factors in regulating animal population numbers.

In contrast, Andrewartha and Birch (1954), following studies of thrips (thunderflies) and grasshoppers in Australia, put forward the view that density-dependent processes are, in general, of minor importance and play no part in determining the abundance of many animal species. They believed that population numbers may be limited in three main ways: (1) by a shortage of material resources, such as food, places in which to make nests, etc.; (2) by the inaccessibility of these material resources relative to the animals' capacities for dispersal and searching; and (3) by a shortage of time when the rate of increase, r, is positive. They felt that the latter was the most important factor determining numbers in nature and that the variation in r would be caused by weather, predators, or any other aspect of the environment which influences the rate of increase. Thus numbers would fluctuate up and down with climatic and environmental variables as r became positive or negative and density-dependence, if present, plays only a minor role in determining population numbers.

These, apparently contradictory, views of population ecology were, with hindsight, simply the result of individuals generalizing from studies of different species in different environments. Andrewartha and Birch's examples were from the variable, semi-arid habitats in southern Australia whereas the later proponents of density-dependence, particularly those who had studied vertebrate populations such as Lack (1966) were looking at more stable environments with more equable climates in temperate regions. It now seems obvious that the position in the range of a species, the type of habitat and environment and the climate must have an effect on the stability or instability of the population and that the 'all-explaining' nature of population theory is a myth. It is important to look at each situation in isolation and evidence of density-dependence may or may not be forthcoming. It is disappointing, but it may be dangerous to generalize for the same species in similar habitats until corroboratory evidence can be produced! However, despite the difficulties of generalization it is clear that density-dependence is acting in many species because many populations are relatively stable and they remain at the level of common, abundant or rare without great fluctuations. Klomp (1962) takes matters further by suggesting that the correlation of population density with climate (e.g. rainfall) indicates regulation by density-dependent factors as without regulation the populations would be found at different densities under similar environmental conditions. This is another way of saying that populations are adapted to their environment.

Adaptation of population density to the environment

Many populations of mammals appear to occur at population densities which are adapted to the environment in which they live; Sinclair (1974d) has shown that buffalo (*Syncerus caffer*) in East Africa have densities

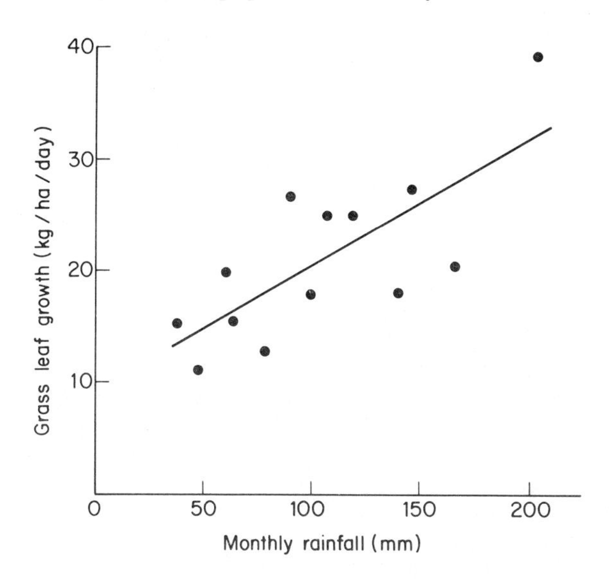

Fig. 6.5 Relationship between the rate of growth of grass leaf eaten by African buffalo (*Syncerus caffer*) and the amount of rain falling in the previous month. Redrawn from Sinclair (1974d).

which are well correlated with the rainfall (equivalent to productivity) of their environment (see Figs 6.5 and 6.6). In evolutionary terms it seems that if populations are stabilized to some extent then this will occur at a level where the risks from overpopulation and starvation are minimized. Assuming that the possibility of dispersal exists (see Chapter 7), it might be just as profitable for an individual to move away from its natal area and have the possibility of finding a more suitable habitat with a lower population density, as it would be to remain in a high density population and run the risk of starvation or being inhibited from breeding successfully or breeding at all. Some populations, as is commonly postulated for deer species such as caribou (*Rangifer tarandus*) (Pimlott, 1967; Bergerud *et al.*, 1983), may have evolved with constant predation pressure preventing catastrophic increases and may now, since man has altered the predator populations by hunting, reach the limit of a resource such as food supply before they show any adaptation of population processes to bring about the limitation of numbers. Other populations, such as many predator species, appear never to reach an excessively high population density and to be very finely 'tuned' to the available resources, so that numbers are limited well below the level set by food or other resources (see Chapter 7).

Density-dependence is an underlying supposition for the population processes of *K*-selected species, but it may be just as important in some relatively *r*-selected species such as many rodents. Studies of red squirrels (*Tamiasciurus hudsonicus*) in Canada (Rusch and Reeder, 1978) indicate that the spring breeding densities are relatively stable despite great variation in autumn densities related to variable mast crops (see Chapter 7). This

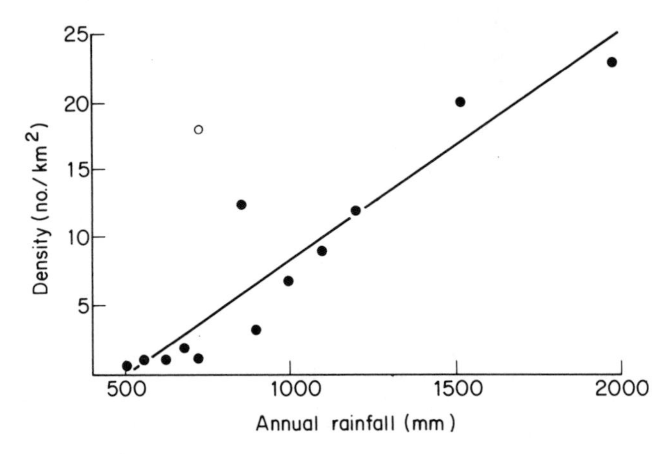

Fig. 6.6 Relationship between the mean crude density of African buffalo in different areas of eastern Africa, including Manyara (○) and the mean annual rainfall for those areas ($P<0.001$). The regression line was calculated for the points with Manyara excluded. Redrawn from Sinclair (1974d).

stable density is interpreted as being the level at which the population can survive even in years of poor cone mast and it operates through density-dependent mortality by territory holders excluding non-territory holders from cone supplies.

Stability in population numbers is interpreted as the result of density-dependent processes acting on the species, but this will not be certain unless movements in one direction or another are counteracted by density-dependent processes, and the population is returned (sooner or later) to the original level. These stabilized population levels may be constantly fluctuating about an equilibrium level rather than following a constant trend, and it may be possible to observe a population which is capable of being stabilized at two or more different levels – a multiple stable state. This is a theoretical situation which has become widely discussed in entomological ecology and although there is still little evidence to show that it is as simple as the model predicts, its relevance to mammal populations is worth consideration. First, however, more general models of population fluctuations and their 'stability domains' must be discussed.

Models of mammalian populations

Theoretical models of mammal populations can become very complex (e.g. Stenseth, 1977), but simple models can give some insight into possible interactions of populations with their environment and still retain some credibility; whether they approach the truth of the situation requires much more research.

Caughley (1976a, b) approaches the modelling of herbivore populations by assuming that their rate of change in numbers is a function of plant density and that the rate of change of the food plants is a function of herbi-

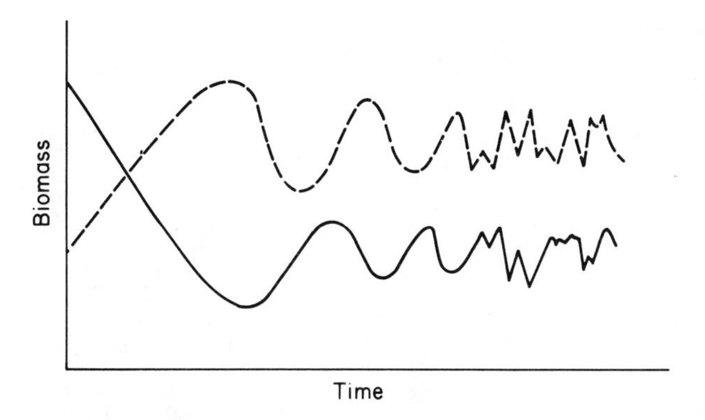

Fig. 6.7 Hypothetical relationships between the plant community (solid line) and a population of herbivores (broken line). Redrawn from Sinclair (1981).

vore density. The two components of the system are assumed to interact as though they were a non-territorial ungulate species, and its food resource. Without going into too much detail of the mathematical assumptions behind the model, Fig. 6.7 shows what might happen under specified conditions of carrying capacity and rate of increase, if a population of herbivores increased after introduction to a new habitat or from a previous reduction in numbers. There is an inverse relationship between both animal and vegetation populations so that both eventually reach an equilibrium point which is relatively stable after some initial sharp fluctuations. Caughley (1976a, b) follows May (1976) and shows that it is equally possible from a similar basic population interaction, but with different defined parameters,

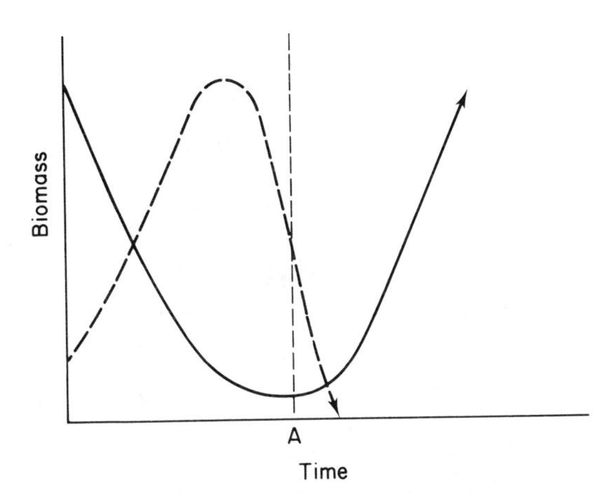

Fig. 6.8 Hypothetical interaction (unstable) between a plant community (solid line) and a population of herbivores (broken line), which becomes extinct. Before time A the behaviour of the system is similar to the early stages of Fig. 6.7. Redrawn from Sinclair (1981).

to have the herbivore population 'trapped' in stable limit cycles (Fig. 6.7) or showing fluctuations where the amplitude is so severe that extinction is likely (Fig. 6.8).

Sinclair (1981) developed these models to explain the various population conditions observed in large herbivores, particularly in relation to 'over-abundance'. Taking the first model above (Fig. 6.7), it is possible that the herbivore population may be perturbed in a downward direction (Fig. 6.9) by the introduction of a predator, possibly by habitat change which would lower the 'carrying capacity', or possibly by loss of habitat. Alternatively the population may be stimulated to increase, after an 'artificial' reduction because of cropping, or after reductions caused by the possibilities outlined above. Whatever the direction of the perturbation, the population will, in this model, reach a new equilibrium with the vegetation and vary again between certain boundaries (Holling, 1973), rather than remain at a steady state with constant numbers from year to year. Scheffer's (1951) study of reindeer (caribou) (*Rangifer tarandus*), introduced onto two of the Pribiloff islands in the Bering Sea in 1911, is interpreted as showing examples of stable equilibria and unstable fluctuations. On St George Island the population grew from 15 to 200 by 1922 and then declined and maintained a level of 40–60; on St Paul Island the population increased from 25 to 2000 by 1938 and then declined to 8 by 1950. This latter fluctuation is interpreted as being similar to the model described by Fig. 6.9. The two islands were ecologically similar and the reasons for the different population states are not certain.

Sinclair suggests that if a population fluctuates between certain boundaries it would be possible to move over an upper or lower boundary and reach a new stable state. The move over a lower boundary could be the result of a catastrophe such as disease, although a catastrophic winter with

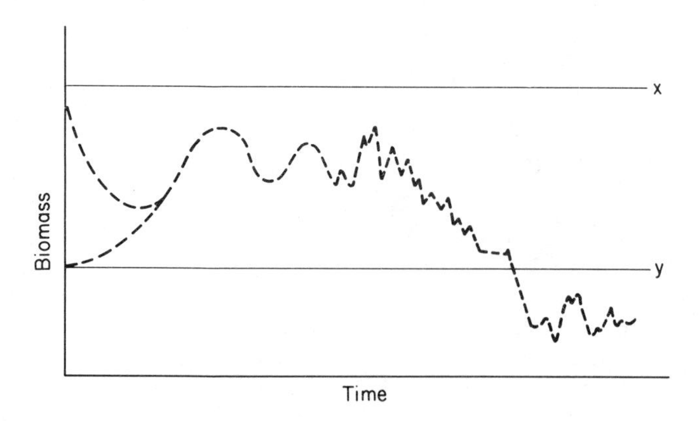

Fig. 6.9 Hypothetical stability domain, bounded by x and y, within which the population or biomass of a herbivore will return to the same equilibrium position. If the population or biomass is forced outside these boundaries a new equilibrium position will be reached. Redrawn from Sinclair (1981).

bad weather could presumably exert the same pressure on the population. A decline and change to a new stable state as a result of disease is thought to have happened in the 19th century in the Serengeti region of Tanzania when the exotic disease rinderpest reduced the numbers of grazing animals (Sinclair, 1981). Subsequently the grazing areas were reduced by the growth of thickets. Similarly, the effect of the disease myxomatosis on rabbits (*Oryctolagus cuniculus*) in Great Britain allowed very significant changes to occur in the vegetation so that much less short-grazed grass was available to the surviving population because the lack of grazing had allowed grassy areas to follow the natural vegetational succession to scrub and woodland or else become very lank grassland (Sumption and Flowerdew, 1985). Consequently the rabbit population has remained below the pre-myxomatosis level although the continued outbreaks of the disease and the possible influence of predators may also be depressing the population. This leads to a further possibility for a change in a stable population level; the effect of a predator on a prey population may be strongly density-dependent at low levels but as prey density increases they are 'released' from predator regulation and the predation becomes inversely density-dependent (also called anti-regulatory). Thus at low densities the predator keeps the prey population at a stable level but as prey reproduction is likely to outstrip predator reproduction, eventually the prey population can escape the effects of the predator. However, it is also possible for the increased intensity of predation on low prey densities to maintain the scarcity of the prey population while the predator switches to additional alternative prey.

The ideas outlined above have parallels in invertebrate ecology (Southwood, 1975; Hassell, 1976) where at low densities predation can keep the prey population at one stable level but if the prey escape this check the population shows the characteristics of an outbreak with high numbers limited by interactions with food supply or some other limiting resource. At this high level the prey population may stabilize or 'crash' to a low level, again depending on the characteristics of the species. The type of population dynamics exhibited by a population could be linked to their position on the '*r-K* continuum' (see p. 102).

There is evidence of more than one stable state occurring in mammal populations. Multiple stable state theory is used in the explanation of the dynamics of populations of moose (*Alces alces*) in 'open' habitats where relatively high wolf:moose ratios exist, and in 'closed' habitats, such as islands, where wolf:moose ratios are lower (Bergerud *et al.*, 1983). The population densities of moose in the open habitat fluctuate around a lower level than those in the closed habitat (Fig. 6.10). It is suggested that moose need more space per individual with higher densities of predation; with relatively more wolves per moose in the open habitat the vulnerable moose cows and calves are widely spaced and kept at the low level to avoid predation. In the closed situation with fewer wolves per moose and more 'escape' habitats such as small islands and shorelines the fluctuations are more marked and around a higher density of moose. However, if wolves are removed from either type of population then it is likely that the moose would increase until limited by some other factor such as food; this upper limit will

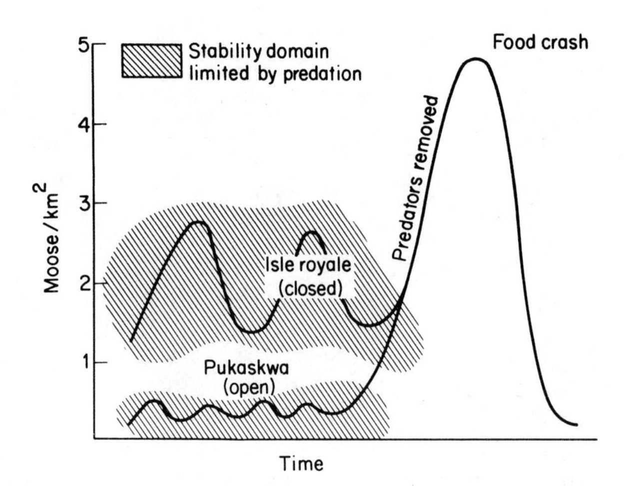

Fig. 6.10 Model of population limitation for moose in the 'open' habitat of Pukaskwa National Park, Ontario, and in the 'closed' habitat of Isle Royale, Michigan. Redrawn from Bergerud *et al.* (1983). © The Wildlife Society.

not be stable but will lead to overuse and major decline.

Sinclair (1981) suggests that another example of a multiple stable state is shown by population changes of the wildebeeste (*Connochaetes taurinus*) in the Kruger National Park, South Africa (Smuts, 1978). The wildebeeste were originally at high density and regulated by interactions with food supply, but game managers decided that there were too many in the population and started a culling programme in 1965. At the same time an increase in rainfall reduced the suitable habitat for wildebeeste by increasing the long grass areas, forcing the population into small groups on areas of shorter grass habitat. This made the population more vulnerable to predation and after 7 years the culling ended; following this the population did not increase as expected but continued in decline because predation was stabilizing numbers at a low level.

Multiple stable states may be possible in many mammal populations and although the obvious examples come mainly from large herbivores, the application of the model to other categories should not be overlooked. Whether the real situation is as simple as the model proposes can only be tested by experiment.

A possible example of multiple stable states from a more 'r'-selected species is seen in the populations of house mice (*Mus musculus*) in Australian wheat fields and the surrounding reed-beds (Newsome, 1969); here predation is not apparently involved but the population is limited at different levels by the resources available. The reed-bed populations colonize the wheat fields at relatively low densities because when the food supply is plentiful in summer the soil is dry and unsuitable for burrowing. However, occasionally rain in summer softens the soil and the population increases rapidly. Newsome (1970) experimented in winter by adding extra wheat to the fields and found that when rain softened the soil and helped burrowing the numbers increased greatly in comparison with control plots where numbers declined because food was limited. The 'plague' induced was uncharacteristic of winter populations of mice and it was concluded that the mice were limited by, and competed for, food and space and that these resources were 'parcelled out' within a social group so that the dominant individuals encouraged the dispersal and reduction of breeding by juveniles. Outbreaks of Mus like this are reviewed by Redhead *et al.* (1985).

The above descriptions of a long-term refuge habitat and the unstable but exploitable man-made habitats come into the realm of Anderson's (1970) ideas of survival and colonizing sites with similar divisions in the population. It may seem extreme to suggest that the genetic composition of the two populations is markedly different but inbreeding in small social units (demes) has been shown in studies of house mice by Anderson and others. Anderson developed these ideas from studies of house mice in small barns which held grain on a farm in Canada; each barn held a social group of about ten weaned mice of which 4–7 were reproductively active and only one or two were males. Immigrant individuals were excluded and so the number in each barn remained stable and excess young were left to colonize other habitats. These barn populations, some less than 1 m apart, were genetically distinct.

Anderson (1970) suggests that not all the individuals in a population behave in the same way and that there will be specific characteristics of survival and colonizing fractions in the population. In survival habitats conditions are stable and the population is in continuous occupancy, inbreeding occurs and the population is stable in structure and density, individuals are long-lived and maturation may be delayed, and selection takes place through differential reproduction determined by a dominance hierarchy. In colonizing habitats conditions are unstable, occupancy is discontinuous, outbreeding occurs, population density and structure are unstable, survival is poor and maturation is rapid, and selection occurs through differential mortality. Thus within one species the characteristics of a population in one area may be very different from those in another. However, it is debatable whether populations in natural situations are as distinct as is claimed for the mice in the barns (see Myers, 1974; Berry 1981).

Population dynamics is thus seen to be more complicated than the description of the numerical changes of a single species in its environment. Both the environment and the species itself are variable and the population may show a number of different dynamic states according to the state of these variables. Exactly how populations of large and small mammals are or are not limited or regulated is the subject of the next chapter.

7

The Ecology of Mammalian Populations

Those species of mammal which are large and easy to observe allow behavioural details such as feeding preferences and social interactions to embellish a study of population dynamics. In contrast, the smaller species are often secretive and nocturnal, and much of what is known of their ecology comes second-hand from live trapping and other indirect methods. To balance this the smaller species generally have shorter life spans, perhaps with the exception of the bats, and so changes in numbers can occur on a much shorter time scale than with large species. It has been argued that elephants (*Loxodonta africana*) fluctuate with a period of 200 years (Caughley, 1976) and although the suggestion has been criticized it is still difficult to test!

Experimental manipulations are relatively easy with small mammals but are now increasingly common in studies of large mammals. In this chapter it is hoped that the reader will be able to gain some understanding of the ways in which mammalian populations are limited or regulated and how closely theories are matched by experimental evidence and description. It is intended that the chapter will highlight the achievements of mammalian population ecologists as well as suggesting areas for further research.

The single- and multi-factorial approach to population ecology

The classical theories of population regulation or non-regulation (Chapter 6) tend to emphasize only a single important factor which influences animal population numbers in a stabilizing or non-stabilizing way. However, these ideas may be too simplistic and the search for a common cause to similar species' population fluctutions or to cyclic fluctuations in mammals may be unrealistic. This is the view put forward by Lidicker (1973) after a long-term study of California voles (*Microtus californicus*) and developed as a general theory (Lidicker, 1978). The model developed from this study envisages a network of environmental factors interacting individually with the subject population and with one another. At any one time one or more of these factors may be predominant in their effects on population density but they cannot be completely extracted from the environmental fabric in which they participate. The model thus rejects unifactor-organism interactions. Both intrinsic and extrinsic factors will be involved in regulation and both vary so that the regulatory machinery is expected to change spatially and temporally. The theory is criticized as untestable because of its vagueness

(Tamarin, 1978b) although Lidicker does emphasize that parts of the model should be tested in isolation for a particular species and a particular population.

Much of the work on population ecology has concentrated on single-factor explanations and this multifactorial approach serves to remind us that the supposed answers to ecological questions may not be as simple as they sometimes look. It is unlikely that a single explanation will cover all events in population ecology except perhaps natural selection! The following sections are largely divided into the main environmental and endogenous factors thought to affect population dynamics in mammals; from the discussion above this division is obviously artificial but serves to underline the importance of these factors in population ecology and, hopefully, their interactions will not be ignored.

The effects of variation in food supply

The food supply available to a population is likely to vary in quality and quantity between seasons and habitats. In addition plant secondary compounds may make what appears to be an adequate food supply unpalatable or indigestible. Thus the assessment of potential food supplies is not easy unless the range of possible food plants taken by the species is known and their seasonal change in quality and quantity as well as their palatability is understood.

To take a simplistic view such as that of Hairston *et al.* (1960), and Slobodkin (1962), the food supply to herbivores is superabundant and so herbivores cannot be limited by food shortage. This statement directed the authors to the conclusion that parasites or predation must be limiting numbers. However, these ideas take no account of competition occurring between individual herbivores and necessarily assume that there is no competition for food. That these ideas are incorrect is now well demonstrated by Sinclair and his co-workers for many African mammalian herbivores and also for many North American species.

In the case of African ungulates the facts are apparently contradictory but further study shows the truth of the matter. Phillipson (1973) estimated the primary production available to large and small mammals in the Serengeti (Tanzania) and came to the conclusion that these mammals took only a small proportion of it so that the total herbivore (invertebrates as well) consumption was only 18% of the annual production. In Sinclair's (1975) study the times of food shortage and abundance are taken into account as well as the annual mean productivity. He considered the different types of herbivore in the Serengeti and came to the conclusion that the herbivores are food limited for part of the year and that this is enough to regulate numbers in a density-dependent manner (see discussion of buffalo p. 14). During the Serengeti dry season (July–November), the mean total requirement for food by the herbivores (large and small mammals and the major invertebrates) in the long grassland, was always in excess of the available food; most of the requirement was due to the ungulates (Fig. 7.1). In the short grassland areas there was virtually no green food in the dry

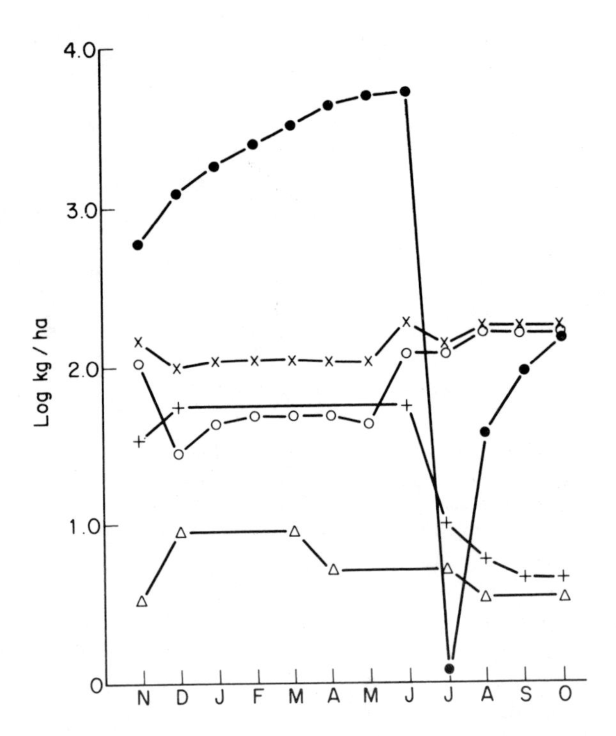

Fig. 7.1 The mean monthly available food and herbivore food requirement in the long grassland in the Serengeti, Tanzania, measured on a log scale. Requirements were in excess of available food for the dry season months July–September but were nearly equal in October. •, available food; ○, ungulate requirement; +, invertebrate requirement; △, small mammal requirement; ×, total herbivore requirement. Redrawn from Sinclair (1975).

season and thus four months of food shortage occurred between July and October. In the small islands of grasses with trees and shrubs there was one month, July, when total herbivore requirements (mostly from small mammals and grasshoppers) were in excess of production and another month, August, when food supply equalled their requirements, giving two months of relative food shortage. The limiting component of the food supply in all cases was the lack of green growth in the dry season. It is advantageous to the long and short grassland habitats that the period when herbivores have the greatest impact on the grass occurs after the time when the grasses have set seed and so growth is finished and the grasses have started to die. Grazing at this time has little impact on the viability of the grasses as they can regenerate quickly and so adjust to the grazing reduction.

Food shortage also occurs in deer populations in North America. However, to take the example of the *Odocoileus* deer species, whether food shortage occurs or not depends upon the type of habitat. In large habitat units showing little diversity, such as the forests of the Great Lake States, dispersal as a population regulatory mechanism is possible only at the

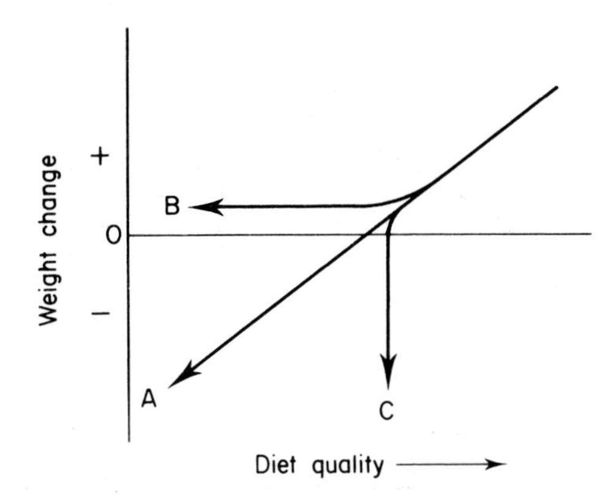

Fig. 7.2 Hypothetical effects on the body weight of snowshoe hares produced by three feeding strategies when the supply of high-quality food declines: relationship A is produced by including low-quality food in the diet but keeping the amount of food eaten per day constant; B by including low-quality food and increasing the daily food consumption; C by eating only high-quality food while allowing daily food consumption to fall. Redrawn from Sinclair *et al.* (1982).

periphery of the habitat units (Klein, 1981); under these conditions it is common for deer to show evidence of malnutrition, reduced fecundity, distorted sex ratios and poor overwinter survival as a result of the high population densities.

Evidence of food shortage, particularly of high quality food, is also seen in the snowshoe hare (*Lepus americanus*) where the response of the hares to declining food supplies was studied in the field and in the laboratory (Sinclair *et al.*, 1982). According to optimal foraging theory (that natural selection should tend to produce individuals which are maximally efficient at propagating their genes and therefore at doing all other activities which ultimately subserve this function; see Krebs and Davies, 1978), when food is abundant animals will select only the better types of food and ignore the others. As foods become scarce more of the lower quality foods will be eaten and so animals should become less selective as food supplies decline. This argument led to the ideas modelled in Fig. 7.2, that as food abundance declines the population will: (1) incorporate lower quality food in their diet and as a result start to lose weight (line A); this assumes that there is some nutritional return over the cost of searching and digesting at a constant rate of intake but with a greater proportion of lower quality food; (2) eat a greater quantity of food and so maintain weight and thus more food would be eaten as quality falls (line B); or (3) limit the diet to high quality foods and reduce the amount of food eaten (line C) and so loss of weight would result. In addition to these diet quality/quantity arguments the point is made that plant secondary compounds may interfere with protein digestion and so increase faecal protein at the expense of digested protein.

Laboratory studies on the hares showed that as food quality declined food intake remained constant and so the poor quality component increased

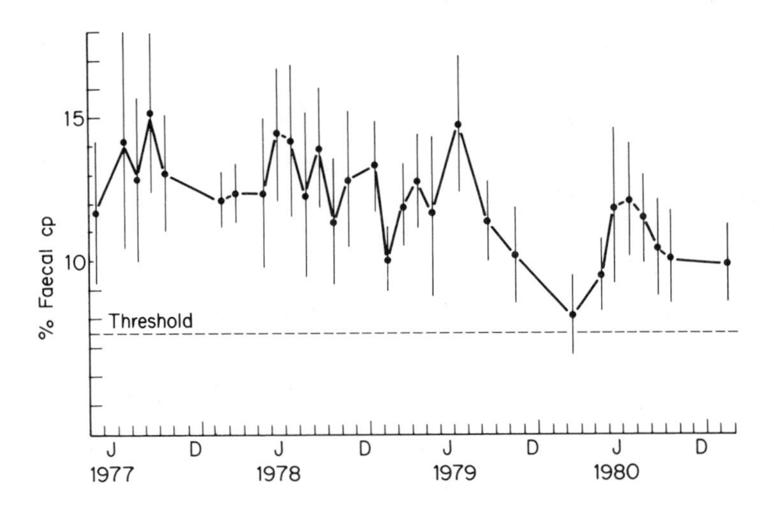

Fig. 7.3 Mean crude protein (cp) in the faeces of snowshoe hares from populations in Yukon. Vertical lines show one standard deviation. Redrawn from Sinclair *et al.* (1982).

and body weight declined as in line A of Fig. 7.2.This means that the quality of the food can be used to predict the ability of the hares to maintain body weight and it was found that the threshold protein quality of the food necessary to maintain body weight was 9% crude protein and that faecal protein was highly correlated with the protein content of the diet. At zero weight gain the diet gave faeces of 7.5% crude protein and this level was used as an indicator of deteriorating food conditions in the field. Samples of faeces from the hare population showed that faecal quality did change with season (Fig. 7.3). The diet changed from herbs in summer to woody stems in winter and the lowest quality food was found to be eaten in the late winter so that some individuals' faecal protein dropped below 7.5%. This indication of a shortage of good quality food and probable weight loss coincided with the time of peak numbers in the population and it is likely that individuals which undergo loss of weight will die; it has been shown that many hares will die if 25% of the body weight is lost (Pease *et al.*, 1979). The results obtained with the snowshoe hares look conservative when the effects of plant secondary compounds are considered. Both phenols and resins in the hares' natural food may hinder digestion but laboratory tests showed that only phenol-impregnated food reduced protein digestion (by 50%). Thus the level of 7.5% crude protein in the faeces in the laboratory may, in the field, be equivalent to a higher level of per cent crude protein if protein digestion is hindered. This would mean that the weight maintenance threshold is at a higher level than 7.5% faecal protein in the field and that an even larger proportion of the population is likely to be losing weight. Further, in the laboratory, the protein level of the food equivalent to zero weight gain was measured at a temperature of 10° C, with no competition for food and no stress from, for instance, food searching, parasites, or reproductive activity. These factors again make the comparisons conservative.

The nutritive value of food is an obvious factor to take into account in studies of food selection but it is also obvious that palatability and the avoidance of plant secondary compounds which may hinder digestion are also important. In a review of this subject as it affects subarctic birds and mammals Bryant and Kuropat (1980) came to the conclusion that the snowshoe hares and mountain hares (*Lepus timidus*) have food preferences which are not controlled by the tissue energy or proximal nutrient content, but probably by the secondary compounds found in the food. Similar conclusions have been drawn for other species including the moose (*Alces alces*): the resinous young growth of the usually preferred food plants such as the Alaska paper birch seems to be particularly indigestible so that moose starve to death despite having rumens full of the twigs of this growth stage (Oldemeyer *et al.*, 1977). The apparent abundance of food but factual limit to food species also applies to some extent to predators: thus the fox (*Vulpes vulpes*) finds common shrews (*Sorex araneus*) distasteful and so a potential food item is not actually eaten (Macdonald, 1977).

The density-dependent regulation of population numbers by food supply appears to occur in buffalo (*Syncerus caffer*) in the grasslands of the Serengeti in East Africa (Sinclair, 1973, 1974, a–d). Studies of the population in different habitats indicated that the amount of rainfall (equivalent to the productivity of the habitat) and the amount of riverine habitat in an area were related to the population density in different areas (see Chapter 6 and Figs 6.5 and 6.6). Undernutrition was evident in juveniles (as well as predation and disease) and was also common in adults of both sexes. The lack of

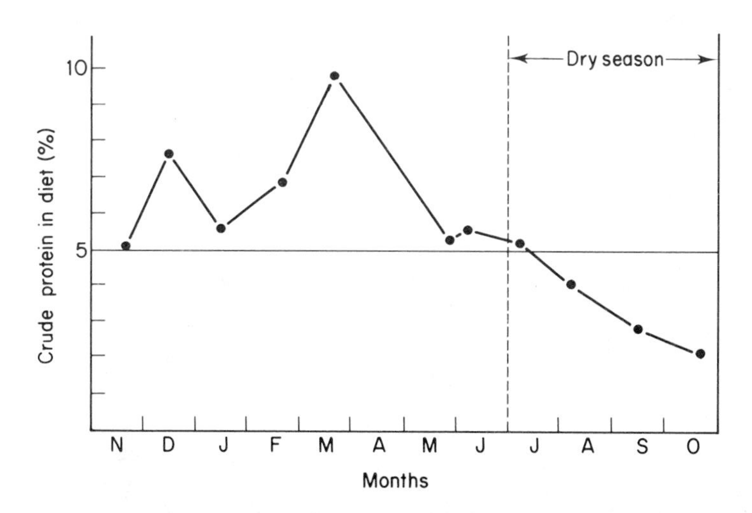

Fig. 7.4 The changes in the percentage of crude protein in the diet of African buffalo in the Serengeti. Note that for most of the dry season the quality of the diet at intake was below the minimum maintenance requirement of 5% crude protein. Redrawn from Sinclair (1974d).

food caused mortality in the dry season when food was in short supply because of the previous grazing of the buffalo and also because of the additional grazing of the wildbeeste (*Connochaetes taurinus*). At this time the buffalo were forced to take an increased proportion of grass stem (high fibre) in their diet and less leaf and leaf base which are both high in nutrients.

During the dry season the buffalos' consumption of crude protein (Fig. 7.4) declined to below 5%, the level estimated to be the minimum maintenance requirement for the species. No evidence is presented to suggest that the poor quality diet affected one class of the population more than another but some individuals seemed to suffer more as a result of the food shortage than others. In the mountainous regions where grass growth was more continuous and more leaf blade was available than in the plains there were higher densities of buffalo which grazed heavily and ate mostly the regrowing leaves; these buffalo also suffered from a lack of available food in the dry season (Fig. 7.5) and as a consequence must have drawn on their body reserves to survive the dry season. In old adults with little fat reserve (partly resulting from worn teeth) the food shortage caused incipient starvation and death which was hastened by predation and disease. All individuals were affected by food shortage but the effects were more severe in the old animals. Intraspecific as well as interspecific competition for food was thus taking place and this cause of adult mortality was able to regulate the population density. That adult mortality due to food shortage was indeed regulating the population is supported by the key-factor analysis (see section on disease). Sinclair points out that if food limitation by itself was the only cause of death there could be a number of buffering

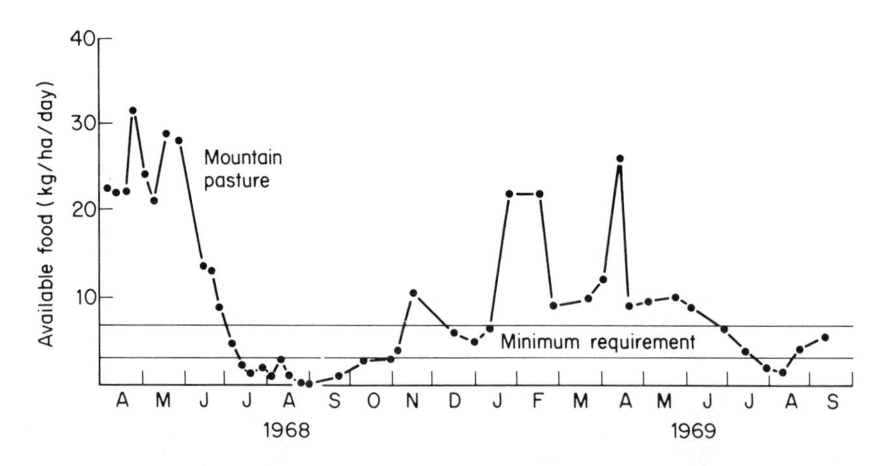

Fig. 7.5 Example of changes in the available food on pastures on Mt Meru, Tanzania. The two horizontal lines enclose the position of the minimum maintenance requirement of the population utilizing those areas. Note that during the dry season the available food was insufficient for the population's needs. Redrawn from Sinclair (1974d).

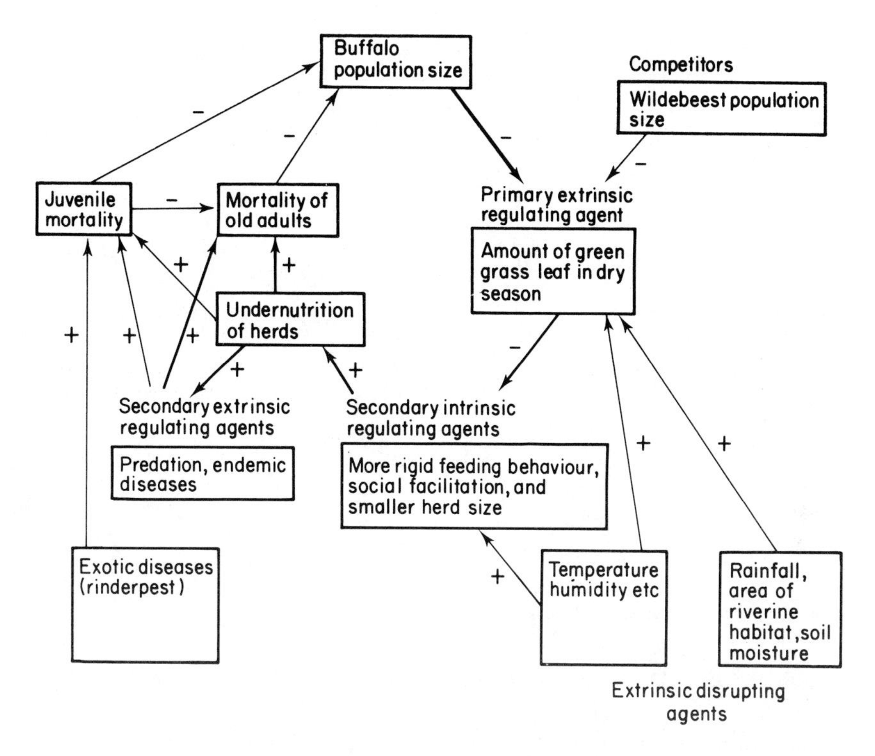

Fig. 7.6 A model of the process of population regulation in the Serengeti buffalo population with the main pathways shown as thicker lines. Minus signs indicate that a higher intensity has negative results, whilst plus signs indicate positive results. Redrawn from Sinclair (1974d).

systems, such as the utilization of fat reserves and migration to more favourable habitats, which might delay the effects of starvation and result in some population 'inertia', allowing a considerable time delay before animals actually succumbed to starvation; this could result in overeating of the food supply and to cyclic fluctuations in both food and buffalo populations (see Chapter 6). However, both disease and predation act as secondary causes of death when buffalo are suffering only moderate under-nutrition, thus reducing the possible time-lag between the depletion of the food and the reaction of the population, making the density-dependent effect of food shortage more immediate, so that the population is highly sensitive to changes in the environment. The juveniles, on the other hand, are so susceptible to diseases, parasites and predators that they suffer large mortalities (mostly from disease) which led to the 'key-factor' disturbance of the population from year to year (see p. 117 for explanation of key-factors). The model proposed by Sinclair to summarize the regulation of the African buffalo populations and the interactions between the environmental factors impinging on the population is shown in Fig. 7.6.

In small mammals the opportunity arises for experimental manipulation

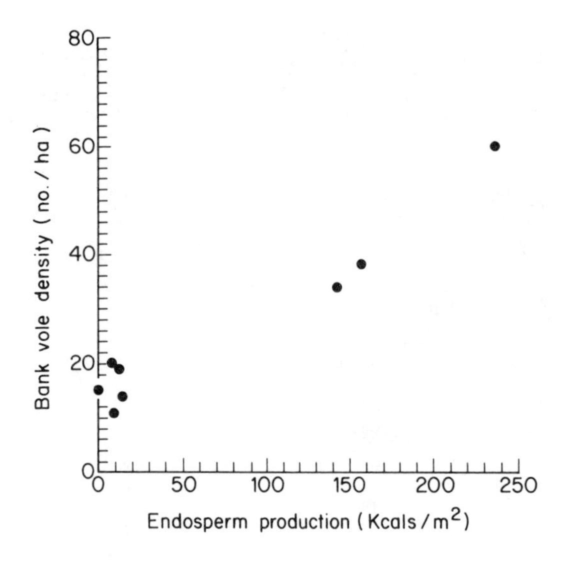

Fig. 7.7 Beech seed production in winter and bank vole density the following autumn. Hestehave Wood, Denmark 1969–78. Redrawn from Jensen (1982).

of populations and their environment as well as descriptive studies showing the effects of food supply on the dynamics of numbers.

It is easy to associate high densities of rodent populations with abundant food supplies but more difficult to assess the availability of food throughout the year, particularly with omnivorous species. Thus in Denmark, the autumn following a winter with a good seed crop of beech mast will generally have high densities of bank voles (*Clethrionomys glareolus*) in beech forest (Fig. 7.7), and poor seed supplies are followed by low densities (Jensen, 1982). On the basis of energy values it seems that small mammals do not eat very much of the potential primary production available to them, rarely exceeding 5% per annum (Table 7.1). However, lemmings (*Lemmus trimucronatus*) at high densities in winter are estimated to be consuming all the grassy (monocotyledonous) stem bases in their habitat between January and the time of the snow melt (Batzli, 1975). Such heavy demands on the food resource fit the descriptions of widespread destruction of habitat and the unusual movements associated with peak lemming populations in Alaska (Thompson, 1955; Pitelka and Schultz, 1964). In deciduous woodland in England it is estimated that wood mice (*Apodemus sylvaticus*) suffer a food limit to population numbers because of a poor acorn crop, on average, in one year out of two (Watts, 1969); wood mice survive poorly into spring when the autumn seed crop is poor (see p. 168 and Fig. 7.20).

Experimental evidence to show that food supply does affect population numbers and survival is available for a number of species. Old-field mice (*Peromyscus polionotus*) were provided with different levels of additional bird seed food; first a moderate level and later a higher level of food. The experimental populations increased but to the same density, with larger

Table 7.1 Energy consumption of small mammals as the percentage of available primary production consumed yearly in temperate ecosystems. After Golley *et al.* (1975).

Ecosystem	Available primary production 10^6 kcal/ha/yr	Percentage consumption
AGRICULTURAL FIELDS		
Rye field	40.7	0.5
Alfalfa	39.8	0.8
Alfalfa	38.8	1.4–21.4
GRASSLANDS		
Grass field	40.6	1.3
Grass field	47.0	1.6
Old-field (seeds only) meadow	0.5	12.0
DESERT		
Desert shrub	2.4	5.5
FOREST PLANTATION	6.7	3.1
FORESTS		
Pine lichen	1.0	1.9
Vaccinium-pine (40 years)	2.4	0.9–1.2
Vaccinium-pine (140 years)	7.0	0.6
Oak-pine	13.0	0.6–0.8
Oak-hornbeam	2.1	4.6
Mixed forest	16.2	0.6
Beech forest	2.0	2.4–3.6
Spruce-grass	14.7–19.2	2.2–3.6
Ciracaceo-Alnetum	–	2.2

differences between experimental and control populations in the winter than the summer (Smith, 1971). It seems possible that some other factor (possibly social behaviour) was limiting numbers at this higher density despite a surplus of food. Similar experimental addition of wheat to a wood mouse population (Flowerdew, 1972) increased summer densities over control densities in one year but not in another (Fig. 7.8); again this may be interpreted as showing that food supply additions only promote increased density up to a limit set by another factor, such as social behaviour, and the lack of any difference between control and experiment in the second year may be simply because both the natural densities and the natural food supplies were as high as they would ever be at that time of year. This idea is taken further by the results of adding supplementary food to a population of white-footed mice (*Peromyscus leucopus*) (Hansen and Batzli, 1979); in this study in North America the additional food had no effect on population density and it was concluded that when natural food was abundant food

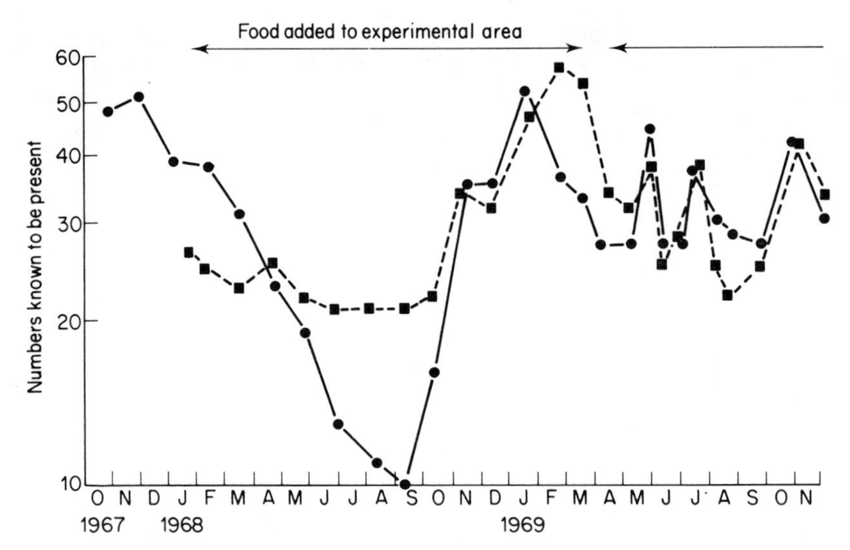

Fig. 7.8 Numbers of wood mice known to be present on the experimental area (▪) and control area (•) in Marley Wood, near Oxford between October 1967 and December 1969. Redrawn from Flowerdew (1972).

availability is not a proximate factor regulating the density of the mice.

In a deer mouse (*Peromyscus maniculatus*) population provided with a supplemental food of whole oats the late winter density increased by immigration to double that of the control population (Taitt, 1981); however, the experimental population still declined into the summer, as these mouse species usually do, and it was thought that behavioural interaction and dispersal (Fairbairn, 1977b) were responsible for this decline. Thus the winter density of the deer mice seems to be limited by the seasonal availability of food but in the spring and summer other factors limit the increase in numbers and help to bring about a decline.

The inability of additional food to prevent declines in numbers is emphasized by two further experiments. In a pioneer experiment of its type, Krebs and DeLong (1965) added fertilizer to a grassland area and also provided a food supplement to the California vole (*Microtus californicus*) population on the area in an attempt to increase food quality as well as quantity. Despite high growth rates and good reproductive rates on the experimental area the voles declined in numbers during the breeding season and had lower rates of survival than the control population. In a later experiment with the same aims (Cole and Batzli, 1978), rabbit pellets were provided for a prairie vole (*Microtus ochrogaster*) population as the pellets were known to be equivalent to a 'high quality' diet from laboratory experiments. In this experiment (Fig. 7.9) the food was added to a poor quality habitat in the hope of increasing vole density and preventing a decline in numbers (commonly seen in voles, see p. 175). During the course of the experiment the voles with extra food reached a higher winter density than the controls (135/ha

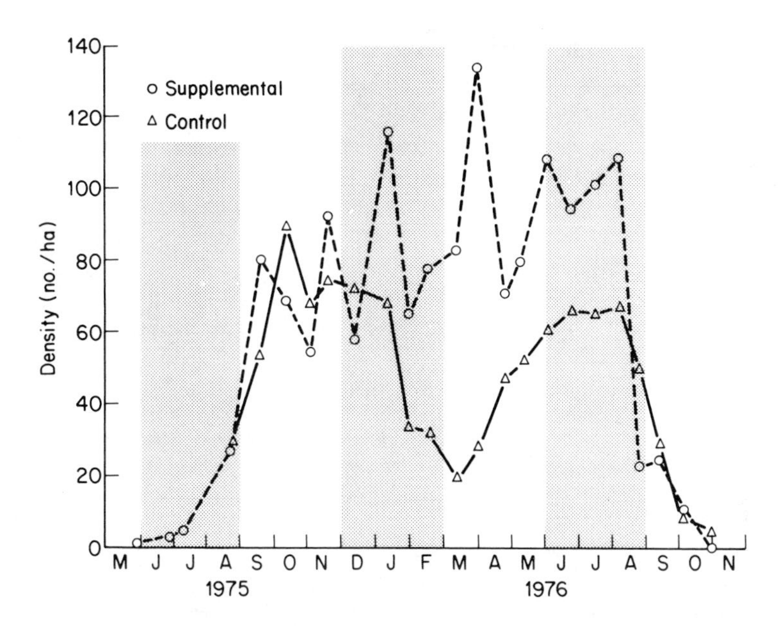

Fig. 7.9 Minimum number of *Microtus ochrogaster* on a trapping grid with supplemental food and on a control grid. Redrawn after Cole and Batzli (1978).

against 90/ha) and showed better body growth and higher litter sizes in the spring and summer. However, both populations declined during the late summer and autumn and so food quality appears to influence the amplitude of the fluctuation in numbers but not the decline. Predation was not thought to be implicated in the decline unless the efficiency of the predators had increased and other factors, perhaps behavioural, were possibly involved.

It may be however, that both food and behaviour are limiting numbers at the same time and this is the conclusion of a further food addition experiment on Townsend voles (*Microtus townsendii*) in Canada (Taitt and Krebs, 1981). High levels of food (laboratory mouse pellets) added to a population led to high densities (734/ha); intermediate levels of food addition led to intermediate densities (504/ha) and low levels of food addition to low density (204/ha). However, as it was also shown that surplus breeding animals are excluded from the population (see p. 162), it seems that both food and behaviour are capable of limiting the number of voles present.

Despite the variation in the results from species to species, these experiments help to show when food limitation is important in the population dynamics of a species and whether food limitation is present at all. The possibility that many species never reach the food limit is brought home by the classic 'fence experiment' of Krebs *et al.* (1969). Here two species of *Microtus* were enclosed in an area of grassland and so not allowed to disperse; these populations did not regulate their densities below the food limit, but overgrazed the habitat and declined in a 'crash' (see Fig. 7.23). A decline associated with obvious food shortage is something that only rarely occurs in these species under normal circumstances, thus the experiment indicates that dispersal is necessary for the normal regulation of numbers in the two species so that the voles compete for space, not food. This topic is discussed further later in this chapter in the consideration of cyclic fluctuations in mammal populations.

Predation

The fact that predation is an obvious mortality factor for many mammalian populations often makes it appear to the lay-man to be an impórtant limitation to population increase. In real life, predators may exert a strong pressure for decrease in mammalian populations but, equally, they may have little effect on numbers. This latter applies as much to herbivores as to

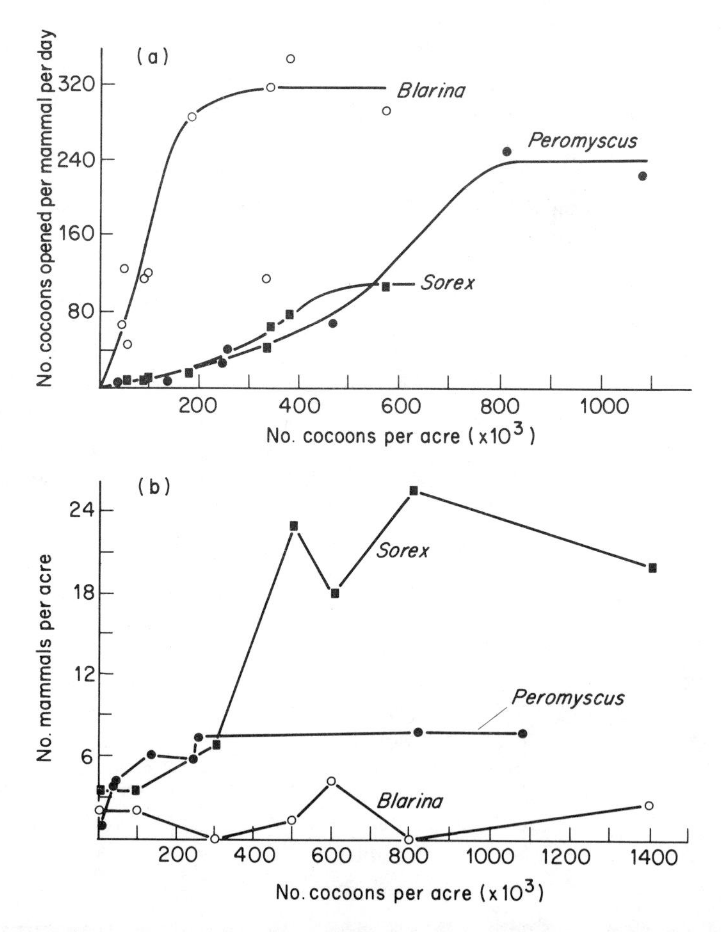

Fig. 7.10 **(a)** Functional responses and **(b)** numerical responses of small mammals to increasing cocoon density. Redrawn from Holling (1959).

species which have few predators capable of taking them as in many large carnivore populations.

Predation is a difficult factor to consider in relation to mammalian population regulation without also considering its interactions with the behaviour and abundance of the prey population and its dependence upon the movements and efficiency, as well as the food preferences, of the predator. Predators may be thought of as 'obligate', feeding on only a few preferred prey or 'facultative', taking whatever species of prey are available at the time. The effects of these various types of predator on the prey population are sometimes very different, as will be seen in the examples below.

Whatever the prey, predators will have a characteristic handling time for each item of prey and this will limit the number of items being taken by one individual predator each day. This characteristic of predators was developed by Holling (1959) and is known as the 'functional response' or the 'behavioural response' which describes how handling time will eventually limit the number of prey taken, whatever the density of prey (Fig. 7.10a). This relationship between prey density and number of prey taken per predator per unit time means that although predators may feed on only one species of prey there is still a limit to the effect of the predator on the prey population. Once the limit has been reached for a particular predator individual the only additional effect the predator can have on the prey is by further predators increasing predator density. A single predator is likely to have a greater effect on the prey population at low prey densities than at high densities and although the interaction with prey may be density-dependent to start with, it is likely to become inversely density-dependent for a range of higher prey densities. If predators do aggregate in areas of high prey density, a phenomenon which is commonly recorded for avian

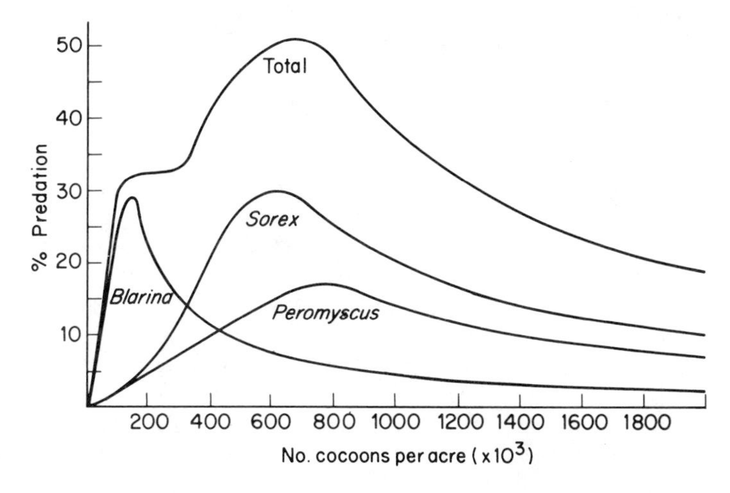

Fig. 7.11 Combined functional and numerical responses to show the relationship between per cent predation and cocoon density. Redrawn from Holling (1959).

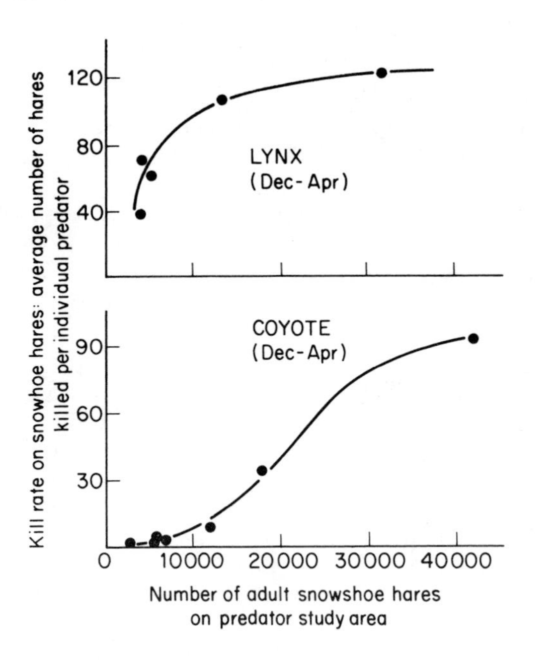

Fig. 7.12 Functional response curves of two snowshoe hare predators at Rochester, Alberta, 1966–75. Redrawn from Keith *et al.* (1977). © The Wildlife Society.

predators, then they conform to Holling's 'numerical response' (Fig. 7.10b). This relationship will show an increase in predator density with prey population density until the predators start to interfere with each other, and so this relationship is also likely to reach an asymptote. If both the functional and numerical responses are combined (Fig. 7.11) then it can be seen that again the total number of prey taken per day by a predator population will eventually reach an upper limit.

Examples of the functional response in mammal populations in Alberta, Canada, are shown (Fig. 7.12) by the lynx (*Lynx canadensis*) and the coyote (*Canis latrans*) which both prey upon the snowshoe hare (*Lepus americanus*). Studies of the diets of these species show that as hare numbers increase the numbers of hares taken per predator reach a limit (Keith *et al.*, 1977). The numerical responses for these and the other predators of snowshoe hares are shown in Fig. 7.13a–c, together with the changes in numbers of snowshoe hares (Fig. 7.13d). In this multi-predator and prey study the snowshoe hares (which show an approximate 10-year cycle in numbers: see p. 184) reached a marked peak in numbers in 1971 with the predators reaching peaks at the same time or later from a direct or delayed density-dependent numerical response. The coyote numbers decreased only moderately in the years immediately after the peak in hares, as did the numbers of great-horned owls (*Bubo virginianus*). The red-tailed hawk (*Buteo jamaicensis*), in contrast to the other species of predator, remained stable in numbers between 1967 and 1975, probably because it has many

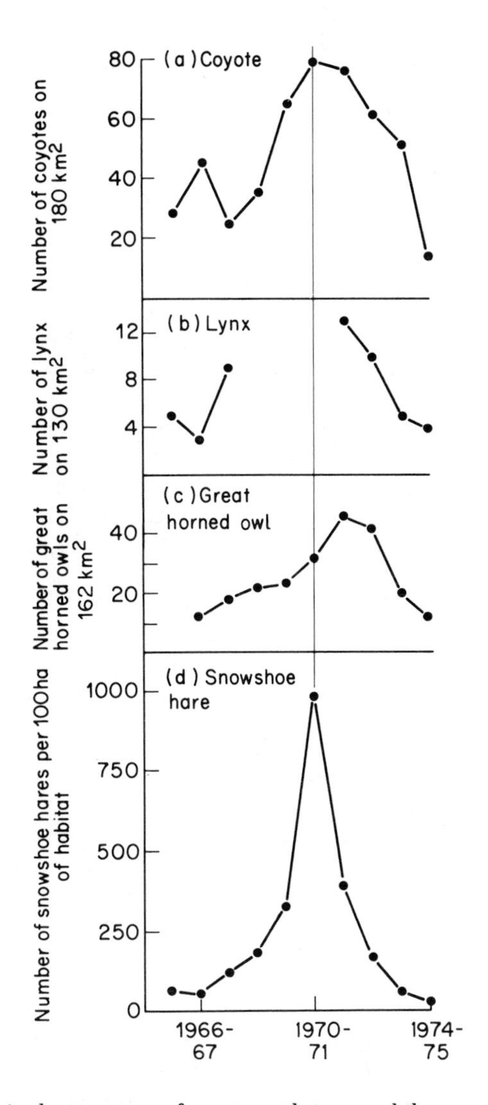

Fig. 7.13 Numerical responses of some predators and the population dynamics of snowshoe hares near Rochester, Alberta over winter. Snowshoe hare densities are for 1 December. Redrawn from Keith *et al.* (1977). © The Wildlife Society.

alternative prey and migrates seasonally to avoid a dependence upon hares in the winter. Despite these predatory losses by the hare population, the peak of the hare numbers was estimated to have lost only 12% to predators, while 43% of the lower density of hares present two years later were lost to predators, causing 22% and 71% of the overwinter mortality respectively. The predators thus seem unimportant mortality agents at peak hare densities but become more important as hare numbers decline, a further instance of inverse density-dependence in the relationship of predation to prey density. Other factors affecting mortality in snowshoe hare cycles are discussed in the later section on cyclic fluctuations.

Inverse density-dependent effects are apparent in the relationships of the tawny owl (*Strix aluco*) with small rodents (the bank vole (*Clethrionomys glareolus*) and probably with the wood mouse (*Apodemus sylvaticus*)) in deciduous woodland in Britain (Southern and Lowe, 1982), for the relationships of predators with the California vole (*Microtus californicus*) (Pearson, 1966), for predators with the black-tailed jackrabbit (*Lepus californicus*) (Wagner and Stoddart, 1972) and for predators with the common vole (*Microtus arvalis*) in Poland (Goszczyński, 1977). These correspond well with the examples given in Chapter 6 of deer populations which are kept in check by predation at low densities, but if they are allowed to 'break out' and reach high densities they are apparently not affected greatly by predation. It seems, however, that predation can act in a density-dependent manner, as seen in the relationship between avian predators and the Townsend vole (*Microtus townsendii*) (Beacham, 1979). In this study in Canada it was found that the mortality of voles attributable to avian predators increased with vole density, though it is not certain if this relationship would continue at higher densities of voles.

The inverse density-dependent action of many predators on prey mortality may be reduced at low prey densities when 'switching' may occur (Murdoch and Oaten, 1975). Switching refers to the change of prey when the usual prey becomes scarce; thus it is likely that the predator will become more efficient in catching and handling a prey type if it comes across it frequently and so the predator will abandon one prey type for another if the latter is more abundant. Weasels (*Mustela nivalis*) in English farmland in years of food (vole) shortage switch to small birds rather than increasing the proportions of all other items in the diet (Tapper, 1979). Similarly male stoats (*Mustela erminea*) in an area of southern Sweden switched from rabbits (*Oryctolagus cuniculus*) to rodents when rabbit numbers declined in 1977 (Erlinge, 1983). Despite the possibility of switching the continued depression of prey populations by predators does occur. This is suggested by studies of 'weasels' (*Mustela erminea*) and (*M. frenata*) preying upon voles (*Microtus montanus*) on Californian mountains, where there are few alternative prey species; the mortality due to *Mustela* predation at low vole densities is probably very great (Fitzgerald, 1977).

Delayed density-dependence is seen in the numerical response of the weasel to the field vole in Tapper's (1979) study in England mentioned above; the weasels react to high vole numbers by increased breeding success and so the numbers of weasels in the year following a high density of voles is also high (Fig. 7.14). In this study it was found that the relative killing rate of voles by the weasels was negatively related to the subsequent changes in vole density and also that the change in weasel numbers after 12 months is negatively related to the change in vole numbers in the following months; thus the weasels may be having an effect on vole population changes as well as voles influencing weasel numbers.

A further study in which a delayed density-dependent effect has been observed is that of the predation of rabbits in an enclosure in New Zealand

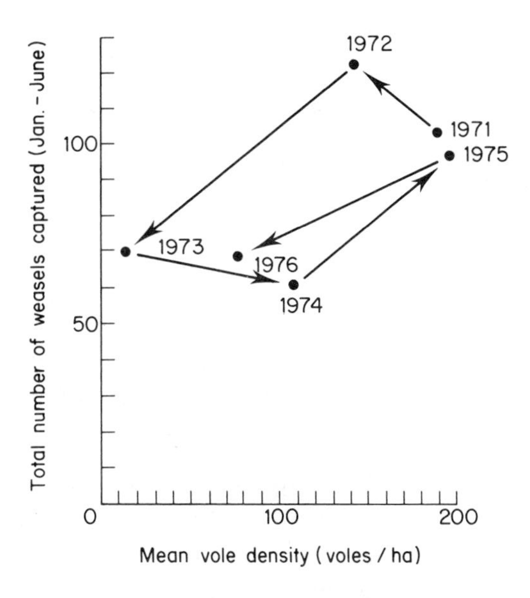

Fig. 7.14 Relationship between the mean density of the field vole (*Microtus agrestis*) populations and the number of weasels (*Mustela nivalis*) captured between January and June on a Hampshire farm each year. Redrawn from Tapper (1979).

(Gibb, 1981). The 8.5 ha enclosure held the rabbit population but allowed the predators (mainly cats and mustelids) free access; this was thus an unnatural situation but it can be argued that more-or-less enclosed populations do occur naturally. It was found that a combination of predation and starvation limited population increase and caused a decline in the rabbits in the 'classical' predator-prey interaction where the delayed density-dependent effects of the predator reduce the peak levels of prey after a time lag (Fig. 7.15). At high densities of rabbits the food supply was overeaten and the susceptibility of the rabbits to predation because of undernutrition and more extensive searching for food led to higher predation and higher numbers of predators entering the enclosure; more rabbits were taken in the diet of the predators than at low rabbit densities. In addition, it is likely that the number of young produced was reduced for much of the time of high density and it is certain that the growth rates of the young were reduced at high density. When all the predators were excluded by a high electric fence the peak population declined (1000 to 13) as a result only of starvation; disease was considered not to be very important, except for coccidiosis which killed many of the young rabbits. The population remained low after the decline for 3½ years owing to the continued presence of greater predation effects (predators were allowed to return) and despite the recovery of the habitat.

Thus in the enclosure many factors were involved in reducing population increase at high density, and if predators were excluded the population reached a food limit where starvation, disease and predation interacted to

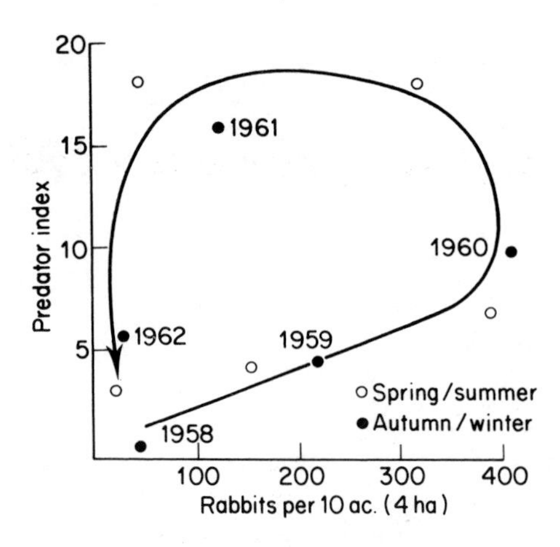

Fig. 7.15 Relationship between the population density of rabbits and an index of the frequency with which feral cats and mustelids were seen in an enclosed study area from 1958 to 1962 (predators were not excluded). Redrawn from Gibb (1981).

reduce numbers. However, the restriction of movement caused by the enclosure must place doubt on the relevance of the results to the natural situation. A high level of predation does occur in some Mediterranean rabbit populations and, with seasonal myxomatosis, appears to limit population increase (Soriguer and Rogers, 1981).

An observational study implicating many predators in the limitation of population numbers has been reported by Erlinge *et al.* (1983). This co-operative study was able to estimate the annual production of the field vole (*Microtus agrestis*) and the wood mouse (*Apodemus sylvaticus*) in the study area and at the same time estimated the food taken by small rodent specialists, the long-eared owl (*Asio otus*), kestrel (*Falco tinnunculus*), the stoat (*Mustela erminea*) and by generalists (facultative predators which included rodent prey in their diet) such as the fox (*Vulpes vulpes*), domestic cat (*Felis catus*), badger (*Meles meles*), polecat (*Mustela putorius*), buzzard (*Buteo buteo*) and the tawny owl (*Strix aluco*). Many of the estimates of predation rates and population density are approximate but, accepting their values, it is concluded that predation by the large number of species is a primary cause of non-cyclicity of these rodent species in the southern Swedish study area, since the annual production of rodents was calculated to be nearly the same as the annual predation on the two species (Table 7.2). This is not found in cyclic populations of rodents when the predators cannot keep up with the rodent production in the increase phase of the cycle (see p. 174). Also, the high predation rate (40–60%) of the voles present at the start of each month in the spring prevented the voles from increasing to very high densities as in cyclic populations. It seems that the wide range of predators were preventing an outbreak of voles by a density-dependent increase in predation; as the rodent populations increased there was a switch in the facultative predators' diets to take advantage of the better rodent food

Table 7.2 Estimated number of field voles and wood mice produced and taken by predators during one year (average of 1975 and 1976). Figures on small mammal consumption by the predators are based on food spectrum, food requirements, and the size of different predator populations at different times of the year. From Erlinge *et al.*(1983).

	No. produced per year	No. consumed by facultative small rodent predators	No. consumed by small rodent specialists	Total no. consumed by predators per year
Field voles	171 400	120 700	36 165	156 865
Wood mice	20 100	17 180	4 366	21 546

supply, particularly by the buzzard, fox and cat. How accurate the estimates really are and how common this type of predator regulation is in rodent populations is open to question. The authors suggest three conditions for the predator regulation of numbers: (1) a rich supply of alternative prey sustaining facultative predators and high predator density, (2) the availability of the small rodents to predation for most of the year, and (3) a heterogeneous environment inducing dispersal and influencing the availability of rodents. Further studies of predation on rodents are awaited with interest.

Turning to larger species of predators and prey, there has been much interest in wolf (*Canis lupus*) predation on deer species in North America. Mech's (1977) observations of wolf predation of white-tailed deer (*Odocoileus virginianus*) showed that the deer which are left after a decline in numbers inhabit the overlapping edges of the wolf pack territories. The wolf packs will only hunt in these boundary areas if they are very short of food because if they do enter the boundary area an often fatal encounter with a neighbouring wolf occurs. Thus the deer survive by avoidance of the main wolf territory and the reduced predation pressure in the boundary areas. In these sanctuaries the deer increase in numbers and colonize the core of the wolf territories with less vulnerable offspring.

Wolves and deer in island situations are often able to interact to maintain relatively stable numbers but extinctions do occur. It is common for a wolf pack to enclose all of the island in their territory and so no reservoirs of deer may be left outside the influence of the wolves; there are several documented examples of deer being exterminated or reduced to very low numbers after the introduction of wolves to previously wolf-free islands in North America (Klein, 1981). An example of a dynamic equilibrium being reached between wolf and prey is seen with the moose (*Alces alces*) on Isle Royale (about 544 km^2) in Lake Superior, also documented by Klein (1981). Moose were present on the island from the early 1900s and wolves reached the island in about 1948. Before the wolves arrived the moose had reached a level of 1000 by 1930 and then as a result of winter starvation and fire a decline to a few hundred individuals occurred in 1948. In the 1960s a stable population of 6000 moose and about 20 wolves was established with fluctuations in numbers related to fire which induced forest succession and so altered the availability of browse for moose. In the killing of moose by the wolves the 'quality' of the moose population was improved because individual moose were tested by the wolves before the kill and those which defended themselves were not attacked. Since the 1970s wolf numbers have declined and then increased with a corresponding increase and decline in the moose population. Recent high numbers (45) of wolves appear to be eating other prey apart from moose.

The examples given above show that the impact of a predator on prey populations depends very much upon the characteristics of the prey population, its habitat preferences, behaviour, and so on, and whether the predator is a specialist or a generalist, whether or not the social structure of the predator allows hunting throughout the habitat and an increase in density to take advantage of an increase in prey density. In addition, the social structure of the prey may determine which individuals form the main proportion of the predator's food.

It has already become obvious that predators are often selective in their predation. This selectivity may have a profound effect on the population dynamics of the prey if breeding individuals are continually removed, but the effect on prey increase may be negligible if animals which are having very little breeding success (see Chapter 2) are taken; the loss of a large proportion of a predominantly non-breeding section of the population will have little effect on the numbers of individuals born into the next generation. This latter assertion is the basis of much sustained-yield harvesting of mammal populations (see Chapter 8). Errington (1946) included the idea of selective predation in his writings on predation and this has had a profound effect on the thinking of ecologists ever since, particularly in relation to the impact of predators on vertebrate populations. Death from predation was viewed as simply a mortality replacing other mortalities which would have happened in any case; thus predation compensated for other mortalities and was of little significance to the population in terms of its effect on long-term dynamics. This interpretation of predation came from studies of populations and assumptions for all populations that numbers were self-regulated and that any 'excess' individuals in the population were a doomed surplus which would die from some mortality agent sooner or later. This idea was of great relevance to the populations studied by Errington, such as the mink (*Mustela vison*) predation upon the muskrat (*Ondatra zibithecus*), but of little use to the understanding of predation in other species systems.

Examples of selective predation have been found in a number of studies of small mammal prey. In a population of Townsend voles (*Microtus townsendii*) in Canada, individuals which were ear-tagged disappeared from the population and were recovered from the pellets of avian predators such as the heron (*Ardea herodius*), and the Marsh hawk (*Circus cynaneus*) (Beecham, 1979). These predators selected males (m:f = 1.49:1) in comparison with the proportion tagged (0.90:1) and also took more small individuals so that small males were the most likely to be caught; these individuals were probably the young males. If males are selected as prey it may be that they are taken because they generally move over greater distances than females and if smaller individuals are taken then it is likely that they are subordinate individuals. The red-tailed hawk (*Buteo jamaicensis*) has been shown to selectively take subordinate cotton rats (*Sigmodon hispidus*), but cats removed the dominant individuals (Roberts and Wolfe, 1974). There are, however, examples of non-selective predation such as that by the tawny owl (*Strix aluco*) on the wood mouse (*Apodemus sylvaticus*) and the bank vole (*Clethrionomys glareolus*) studied near Oxford by Southern and Lowe (1982). Here the mice carried numbered metal rings (not now recommended) and the recovery of the rings in the pellets (castings) of the owls indicated no selection by sex or weight.

Experimental studies of predation are difficult to carry out and thus are seldom reported; two such studies of predation on populations of the cotton rat (*Sigmodon hispidus*) (Schnell, 1968; Wiegert, 1972) help to show how predation affects the population dynamics and at which densities it is important in the ecology of the rats. In the earlier study of the cotton rats in enclosures, a high density population which was protected by netting from avian predators declined only slowly compared with a sharp decrease in a control population, but when both populations reached a level of 22–37 per hectare they remained at this level. Low density populations showed little difference in the low mortality rates whether predation was allowed or not. Similarly, in Wiegert's experiments in similar enclosures the decline from high density over the autumn and winter showed reduced mortality with protection from predators and remained similar to the 'wild' situation when predation was allowed. In this experiment the peak density was maintained long after the mortality due to predators would normally have reduced the population to a low level. It thus seems that the avian predators help to reduce the population from a high density to a lower density but no further. Whether these results would be the same if dispersal was allowed is still to be tested, and it is not stated which other prey species are turned to at low rat densities.

Thus the effects of predation on a population, both quantitatively and qualitatively, are variable within and between species. Undoubtedly large mortalities can be inflicted by predators but it seems unlikely that many mammal species will be regulated by predation, rather the dynamics of the predator population are likely to be strongly influenced by the dynamics of the prey.

Disease

The influence of disease, including infection by parasites, on mammalian population dynamics has been badly neglected and is often overlooked. The theoretical approach to the effect of disease and parasites on the population dynamics of animals has been recently developed by Anderson and May (1979; May and Anderson, 1979). Their definition of parasites includes viruses, protozoa, bacteria, helminths and arthropods and they suggest that these factors are likely to play a part analogous, or at least complementary, to that of predators or resource limitation in constraining the growth of animal (and plant) populations. There is still little evidence on which to test this idea but it should be born in mind in any population study.

Anderson and May suggest that parasites can influence the region of the 'dynamical landscape' in which a stable state is maintained by a number of different mechanisms. It would be possible for a disease not to be maintained if the rate of transmission was low but, above a certain threshold, it would be endemic and, for example, in schistosomiasis (a parasitic disease of man) worm pairing for sexual reproduction in the human (primary) host would continue to maintain the infection in the population. Another possibility is the effect of the nutritional state of the host on the parasite pathogenicity; malnourished hosts have a lowered immunological competence and are less able to withstand infections. Thus the pathogenicity of a parasite will increase as host density rises to a level where competition for food is severe. This might lead to two stable states, one with high host density and low 'worm burdens' and the other, depressed by the disease to a low host density and a high or average burden of the parasites. High death rates are considered to be common in high populations because of parasite pathogenicity related to increased levels of stress (see p. 178) rather than by increased levels of transmission as envisaged by Lack (1966), though this may vary with the parasite.

The regulatory potential of parasites is considered to change with evolution. A parasite may regulate a host population at a stable equilibrium during their early association but, as selection pressure reduces the average susceptibility of the hosts, the regulatory effect will tend to wane and eventually the host may escape being regulated by the parasite infection. Much more study of host–parasite interactions in the field is needed. Parasites under this broad definition need not necessarily regulate population numbers, but they can act as powerful agents for decline and become key-factors.

The action of disease at high density is seen in Gibb's (1981) enclosed rabbit population in New Zealand where coccidiosis killed many young rabbits. A further example from rabbits is that latent myxomatosis can become lethal when the population is stressed (see p. 178) (Myers, 1971). Myxomatosis in rabbits in Britain and elsewhere (e.g. the south of France) is a regular occurrence, often taking many juveniles in summer and causing a marked annual fluctuation in numbers (Vaughan and Vaughan, 1968; Rogers, 1981). However, the disease is not the only cause of high juvenile mortality and predation also has a marked effect on the rate of population increase (Lloyd, 1981).

A strong density-independent effect of disease in reducing the population density is described by Sinclair (1973; 1974c) for the African buffalo (*Syncerus caffer*) population in the Serengeti (see section on food supply). Key-factor analysis (see p. 117) shows (Fig. 7.16) that the key-factor contributing on average the most to the annual mortality (K) was k_j, the juvenile mortality, and that this was most likely to displace the population from its equilibrium level, causing short-term fluctuations from year to year. Juveniles are more susceptible than adults to epidemic diseases such as rinderpest (a disease related to measles introduced into African wildlife from domestic cattle) as well as to predation. Rinderpest caused epidemics in the 1950s in the Serengeti but it had disappeared from the buffalo between

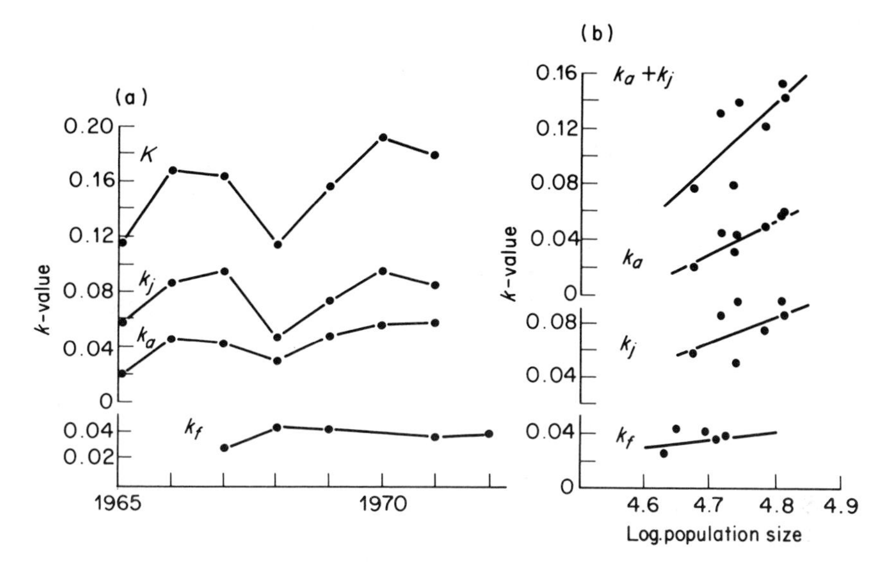

Fig. 7.16 **(a)** The total annual reduction (K) and the separate k values plotted consecutively for a population of African buffalo. Juvenile mortality (k_j) contributes most to the fluctuations in K, and is the key-factor (see text for further explanation). **(b)** The log reduction (k value) at each stage of the annual population cycle of African buffalo plotted against population size prior to that reduction; k_f represents the reduction in fertility, k_j juvenile mortality, k_a adult mortality, and $k_a + k_j$ the combined adult and juvenile mortality. Only the regression line for k_a has a slope which is significantly different from zero at $P<0.05$, suggesting density-dependence. Redrawn after Sinclair (1973).

1962 and 1963. Sinclair developed models for buffalo population dynamics, using the partitioned mortalities from his key-factor analysis, which are consistent with the disappearance of the disease at this time and which suggest that rinderpest mortalities in the past reduced the rate at which the population was increasing to the 1970s equilibrium level. The reduction in the mortality due to rinderpest allowed a more rapid increase to this level after the early 1960s.

The models of buffalo dynamics (Fig. 7.17) use the three k values (Fig. 7.16), of which only k_a the adult mortality, is shown to be density-dependent. Various combinations of supposedly density-dependent k values and average values for the others were used to simulate the population increase from the start of Sinclair's work in 1965 to 1972, and also to simulate the previous change in numbers backwards from 1965 to the late 1950s. Model I had density-dependent adult mortality and average figures for juvenile mortality and reduction in fertility; Model II had adult and juvenile mortality density-dependent and average figures for reduction in fertility; and Model III had fertility loss density-dependent with average figures for the adult and juvenile mortality. Models I and II produced the best fits to the actual population counts from 1965 to 1972 (Fig. 7.17b) but

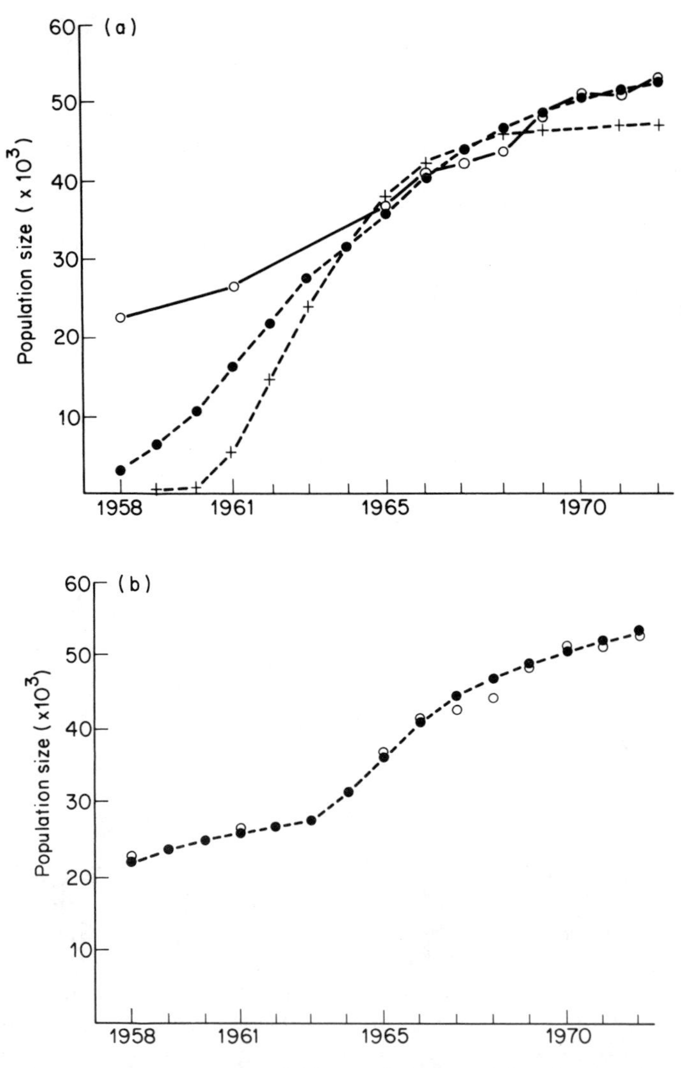

Fig. 7.17 **(a)** The census figures of African buffalo adults plus yearlings (○) compared with two model populations. Model 1 (•) has only adult mortality regulating. Model 2 has combined adult and juvenile mortalities regulating (+). **(b)** The observed census data of African buffalo (○) compared with the reconstructed population using Model 1 for the years 1964–72, and a modified Model 1 simulating the effects of rinderpest for the years 1958–63 (•). Redrawn from Sinclair (1973).

did not mimic the changes before 1965, showing a faster increase than was observed. Sinclair argues that the conditions acting on the population were different before 1965 compared to those after that date, and that these conditions must have changed at some time between 1961 and 1965 since the models do not fit the 1961 figure; the changed conditions before 1965 must have resulted in a reduction in fertility or an increase in mortality, compared with the average after 1965, in order to have produced the slower increase in buffalo numbers. The buffalo population must have been displaced from its equilibrium level before 1965 and was therefore increasing towards it during the 1960s. Since it was known that rinderpest disappeared between 1962 and 1963, this fact was used to modify Model I so that the population was estimated in the normal way from 1965 but that before 1964 Model I had a higher constant juvenile mortality for the buffalo calves. Although figures for the effect of rinderpest on buffalo populations were lacking, it was known that wildebeeste (*Connochaetes taurinus*) showed similar effects in the same area, so that the mortality of calves when rinderpest was present was greater than mortality without rinderpest by factor of $\times 1.76$. Thus the key-factor juvenile mortality was increased by this factor in Model I for the earlier years but other reductions were kept as before. The extrapolation back from 1964 to 1958 approximated well to the observed census figures for 1958 and 1961 (Fig. 7.17) and thus there is good circumstantial evidence that rinderpest was responsible for the generally low population levels of buffalo in the 1950s and early 1960s.

Disease has caused dramatic changes in other mammalian populations, such as the decline in the rabbit in Europe after myxomatosis, from which the population has still not recovered. It is possible that the rabbit in Great Britain will never become as numerous or as widespread as it used to be because the habitat has changed and much of the previously favourable grassland has become lank or transformed into scrub or woodland, and because the predator population is probably greater now than before myxomatosis.

The effects of a parasite of one species on the population dynamics of another species can also be dramatic. After myxomatosis the stomach worm (*Graphidium strigosum*) of the rabbit (to which the rabbit is well adapted) was probably less easily transmitted to the hare (*Lepus europaeus*) because fewer were shed in the faeces of the reduced rabbit population, and Broekhuizen (1975) suggests that this is a cause of the increase in the hare population since myxomatosis. However, there are other factors such as habitat changes and lack of competition to consider. In moose (*Alces alces*) in Nova Scotia the population was greatly reduced after 1940 through infection by a nematode parasite causing cerebrospinal nematodiasis. This nematode is carried by the white-tailed deer (*Odocoileus virginianus*), which is better adapted to the worm, and now the ranges of the two species overlap very little (Telfer, 1967).

Behavioural factors affecting mammal population dynamics

Behaviour as the external expression of the interaction between physiology and environment is an obvious mediator of cohesion or dispersion within a population. Exactly how individuals within a population behave will depend on their social environment and their innate responses as well as their learned behaviour patterns and the stimuli in their immediate environment.

The early part of the life is, for mammals, commonly spent with the mother and this dependency forms the basis of many social organizations which develop in populations. After weaning individuals of many species form single sex groups, in others they are solitary and in some they take their place in an established social organization. Within this structure, if it exists, individuals often form a dominance hierarchy, with dominant and subordinate individuals, and their place in this hierarchy can have a marked effect on their breeding success as discussed earlier (p. 42 and Table 2.2). Further examples of social restriction affecting breeding success can be found for house mice in studies documented by Crowcroft and Rowe (1963) and for red deer by Clutton-Brock *et al.* (1982).

If social behaviour structures the population and determines which individuals shall live where, then it is a potent force in the population ecology of the species because spacing behaviour has the potential of limiting the numbers of individuals inhabiting a unit area of habitat. Such behaviour includes hierarchical behaviour which apart from having the potential of limiting breeding success may encourage dispersal and so limit numbers; home range (the area traversed by an individual in its normal activities of food gathering, mating and caring for the young (Burt, 1943)) behaviour, which is often linked with hierarchical behaviour, will also have the potential to limit numbers although much overlapping of movements and much tolerance of conspecifics is commonly observed; and territorial behaviour which is equated with mutual exclusion from an area so that neighbours are very intolerant of each other and little overlapping of movement occurs once territories have been taken up. Intruders into territories are usually forcefully ejected by the territory owner and once this has been reinforced a number of times the situation becomes relatively stable. This latter form of spacing behaviour is often assumed to limited population numbers but the size of territories may be linked with food supply and so caution must be exercised in interpreting the behavioural information; territories have other functions besides the possible limitation of population numbers, for example the maximizing of food supples and the attraction of a mate.

Social behaviour is thus inextricably linked with population density and sometimes with population limitation, the cohesive or dispersive nature of the behaviour determining the density of the population.

The demonstration of the behavioural limitation of numbers

In order to show that behavioural factors, via socially induced mortality, or depression of recruitment, limit a breeding population a number of conditions must be met. These were first put forward by Watson and Moss (1970) and have recently been restated by Taitt (1981) and Moss *et al.* (1982). These conditions are as follows.

(a) A substantial part of the population does not breed. This may happen as a result of deaths of potential breeders, by individuals attempting to breed but without success because they and/or their young die, or because some individuals are inhibited from breeding at that time.

(b) Such non-breeders are physiologically capable of breeding if the dominant or territorial (breeding) individuals are removed.
(c) The breeding population is not completely using up a resource such as food, space or nest sites. The resource itself is limiting if this is so.
(d) The mortality or depressed recruitment due to the limiting factor(s) changes in an opposite sense to compensate for other causes of mortality or depressed recruitment. Thus limitation by spacing behaviour will relax after heavy losses from other causes and will become more intense if the other mortality or depressed recruitment is low.
(e) If conditions (a)–(d) are met, and numbers change following changes in food, then food and behaviour are both limiting population size.

In using the word 'limiting' rather than 'regulating' we are accepting the argument that density-dependence may or may not be involved in determining population numbers. If density-dependence is observed in a limiting factor then the population is likely to be regulated. If, however, the limiting factor acts at a threshold density above which no further increase takes place then we know that density-dependence is not involved. Limiting factors may thus work in a density-dependent manner or act in an all-or-nothing manner at the threshold so that the population growth suppressing effects increase suddenly in their intensity.

In field studies of mammals it is unusual to know if any or all of the above conditions are being fulfilled, but some progress has been made, especially with small rodents. Studies of other types of mammals often show behavioural limitation of numbers but do not necessarily refer to Watson and Moss's criteria or consider what experimental evidence is needed to confirm the observations. It is commonly proposed that food supply or some facet of the environment sets the ultimate limit to population growth but that social organization will provide a proximate limitation, and also that this has evolved to maximize individual fitness (e.g. Tamarin, 1983). It must also be accepted, as with the functions of territory, that it may be simply that spacing behaviour spaces the population out once numbers have been limited by starvation or some other factor (Lack, 1966). Which situation is true must depend on careful observation and experiment.

Social and spatial organization affecting population numbers

Watson and Moss's criteria apply to the breeding population as this is the part of the population which is functioning from one year to the next and which lays the foundation for the determination of how many individuals will be present in the future. However, it is possible within this breeding structure for the male population to be limited in a different way to the female population and this should always be kept in mind. In the following section a number of examples of apparent population limitation by social behaviour are outlined.

Non-breeding animals are relatively easy to find in some species of large mammals and are commonly found in, for instance, the roe deer (*Capreolus capreolus*). The dominant bucks of this species occupy territories and evict

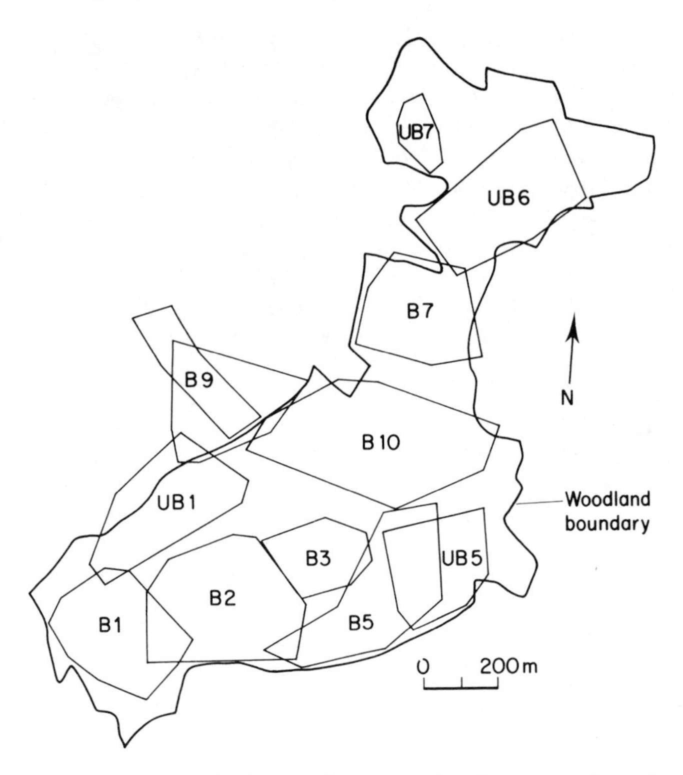

Fig. 7.18 Territories of marked (B) and unmarked (UB) territorial roe deer bucks in 1968. Redrawn from Bramley (1970).

younger bucks which do not breed, do not survive as well as the dominants and emigrate more. Bramley (1970) found that a 113 ha wood supported eleven territorial male roe bucks (Fig. 7.18) and that their number remained constant for the three years of his study. In one year there were 11 territorial bucks, 12 non-territorial bucks, 21 adult females, and 12 male and 5 female yearlings. The bucks without territories seemed to be the younger males and their range of movement was greater than that of the territorial bucks, probably leading to dispersal by the next year when none were seen again. The female home ranges overlapped each other and the ranges of the bucks.

In the timber wolf (*Canis lupus*) which inhabits large areas of North America the populations are broken up into packs and their biology has been reviewed by Packard and Mech (1982). Members of the pack are ranked in an hierarchical manner and form a closely related family group. The pack maintains a territory so that the density of packs (breeding units) is limited. Usually, only one female breeds and the non-breeding pack members help the breeders in hunting and the care of the young. Mating takes place in February or March and after a two-month pregnancy five or six pups are born. Given good feeding conditions, the pups are fully grown by autumn and accompany the adults during the winter. Some young remain with the adults as successive litters are born but many die or disperse. The dispersers contribute to a surplus non-breeding population and wander over large areas between territories, avoiding the packs. The mortality rate of these lone wolves is high and they do not reproduce unless they acquire a territory as well as a mate. The effect of social, maturational and nutritional factors on dispersal is unknown. Wolf populations often appear stable and the social limitation of numbers by territoriality appears to be strong, so that the growth of packs is regulated by social bonds and competition for food, and the growth of the breeding units is regulated by territoriality and prey density (Mech, 1970). However, the 'stable' density does react to food availability and so the saturation density is not independent of food resources; abundant food supplies affect wolf numbers in the long term so that the numerical responses of wolves show a time lag of a few years. Social factors directly affect the population size by intraspecific strife and the limitation of the number of breeding females and indirectly by their interaction with nutritional factors, so determining which segment of the population will be most influenced by a change in food supply. It is possible under abundant food conditions for an extra female to breed and raise her young successfully.

Population regulation in the mountain lion (*Puma concolor*) has been studied by Horncocker (1969, 1970) in Idaho, North America, where there is an abundance of prey. It was possible to divide the population into resident adult males, adult females, and adult females with young up to two years old. Young lions disperse at the age of 2 years and some young are killed by adults; transient individuals, mainly young adult males and females, would occasionally be seen passing through the study areas. The resident lions are territorial and the territories of adjacent adult males did not overlap each other but did overlap the territories of adult females. The female 'territories' were not as discrete as those of the males and overlapped each other to some extent. Avoidance behaviour allowed individuals to separate themselves in time and space and so avoid overt aggression between individuals, and although transient animals may enter an occupied area with impunity, they only remain if the resident dominant male has died.

In the African lion (*Panthera leo*) Bertram (1973) found that the regulation of numbers was similar to that of the mountain lion. The territory size of the pride is relatively constant and the pride will consist of 2–4 adult males, 4–10 breeding females and their offspring. Nomadic non-breeding males and non-breeding females are present in the general area. Pride size, in terms of the number of adult females, does not change in response to short-term changes in food availability and seems to be regulated in a density-dependent manner. Despite the high cub mortality (cubs are last in the dominance hierarchy at small kills and often starve to death), the more than adequate reproductive rate is able to replace the losses of adult females. Bertram concludes that condition (e) of Watson and Moss (see p. 163) is fulfilled as both food and behaviour are limiting numbers. A proportion of subadult females leave or are driven out of the pride at about 3 years of age and all subadult males are ejected at this age. Thus pride size in terms of adult females increases little and more females are probably evicted at higher pride sizes; density-dependent recruitment or eviction being the mechanism involved in the regulation of population numbers, but more information is still needed. Expelled females have only a small chance of breeding and join the nomadic population; they will not usually be allowed back into the pride territory.

In trying to discover how rabbit (*Oryctolagus cuniculus*) populations are regulated, if at all, the following examples indicate that different mechanisms may be operating in different geographical locations. Henderson (1979, 1981), working on the Scottish coast, has shown that increased food supply increases the production of young per female, but that increased dispersal from high food quality areas allows population density to remain similar to areas without additional food. The number of breeding adults is determined in the late winter by intraspecific aggression; enclosure studies show that the individuals which die are less dominant and become more scarred. Rabbit corpses were easy to collect and it was suggested that most adults die *in situ* during the spring reorganization period and that the surviving juveniles disperse during this period. Studies of rabbit populations in large enclosures in Australia (Mykytowycz and Fullagar, 1973) show that breeding success is very variable (Table 2.2) and that the young of subordinate females, if they are produced at all, suffer greater mortality than the young of dominant females. These results from enclosures and free-living populations are interpreted as showing a strong social influence on population increase, but it is also suggested that to be effective environmental factors must also be involved. In arid areas the role of climatic conditions is shown in the regular summer mortality of young rabbits due to starvation when plant food rapidly loses quality in high temperatures. The flooding of burrows, predation and coccidiosis also contribute to mortality, as well, of course, as myxomatosis. It is considered that the social organization of the rabbit has become adjusted to environmental conditions characteristic of the Mediterranean region where the species evolved, and that only with the assistance of certain extrinsic factors is it capable of curbing the increase in numbers. Changes in any environmental factor such as prolonged favourable weather, the absence of predators, or changes in plant cover may weaken the effectiveness of the social influence on numbers and allow the population to expand. Whether social influences on population numbers are more intense in Scotland than in the Mediterranean or Australia is still open to question. In other studies of rabbits in Australia (Wood, 1980), predation is shown to have more impact on numbers than the climatic effects. This occurs in the sandy areas of the less harsh arid environment, where numbers are relatively stable from one year to the next, and rabbits are able to survive the periodic droughts. These studies emphasize different aspects of the rabbits' environment as being important in stabilizing numbers or allowing numbers to be stabilized by social behaviour; which situation is most common in rabbit populations is a topic for further research.

The Arctic ground squirrel (*Spermophilus undulatus*) shows strong territorial limitation of the breeding population (Carl, 1971). In the study area near Point Barrow, Alaska, the breeding population remained relatively constant in numbers from one year to the next during three years, and territory ownership was observed to be contested by males. Territory ownership stabilizes the numbers of breeding individuals each year and only those with territories at the start of the breeding season breed. Females set up breeding territories in association with a male and surplus females emigrate before the breeding season. The surplus, non-breeding, males are of all ages and make up about half of the male population in the area. The surplus males exist as 'floaters' in the breeding colonies and they do not occupy defended territories unless one is taken over when its owner dies. The surplus squirrels are expelled from the main colony and are more likely to be killed by flooding or by predation by foxes or bears. Juvenile ground squirrels grow during the summer, and further contests occur for hibernation sites after the original territorial structure has broken up; many more individuals find hibernation sites than breeding territories, and so numbers are higher in the study area in the winter than in the summer.

In many rodent species where the observation of social behaviour is

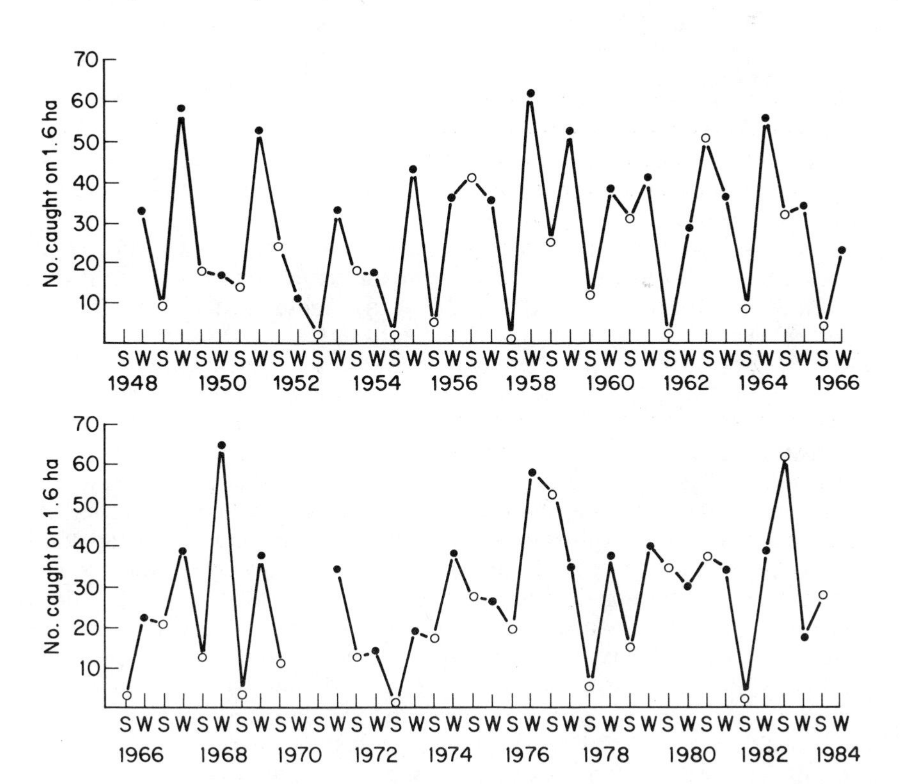

Fig. 7.19 Trap catches of wood mice at six-monthly intervals (S = summer, May–June; W = winter, November–December) on 1.6 ha of Wytham Great Wood, near Oxford. Redrawn from Flowerdew (1985) after Southern and Lowe (1982) with additions. (H.N. Southern and K. Marsland, personal communication). By permission of the Zoological Society of London.

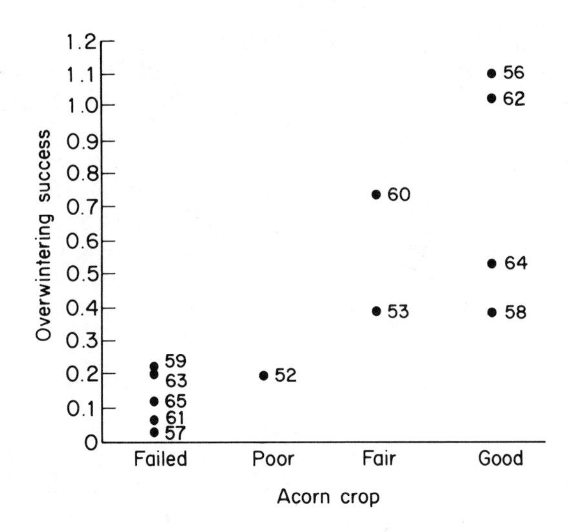

Fig. 7.20 The overwintering success (June adults as a proportion of the total population in December) of wood mice in relation to the acorn crop the previous autumn. Numbers indicate December data. Redrawn from Watts (1969).

difficult, such as in the wood mouse (*Apodemus sylvaticus*), the demonstration of population regulation by social behaviour relies on indirect methods of study.

In the winter the numbers of the wood mouse in Wytham Woods, near Oxford, are more constant than in the summer, varying between about 10 to 66 from one winter to the next (Fig. 7.19). The numbers in summer, although often constant for a number of months during each year, are more variable having been between 2 and 55 on the same area of woodland during the past 36 years (Fig. 7.19). This is thought to be because summer numbers are largely determined by the size of the previous autumn's acorn crop and so overwinter survival varies greatly with the size of the acorn crop (Watts, 1969) (see Fig. 7.20). Thus within each year numbers fluctuate from winter to summer, usually declining markedly at the start of the breeding season and then increasing with the recruitment of many juveniles at the end of the breeding season. Although juveniles are produced throughout the summer they are not usually recruited and the survival of the early cohorts is poor. Only when spring and early summer densities are very low do numbers increase immediately and rise to the winter peak with no summer check. At the start of the increase in numbers the density doubles each month and exactly when this increase starts is related to the density in the summer. Watts (1969) shows that this density-dependent relationship between the start of the increase and summer numbers allows relatively constant winter densities to be achieved despite very variable summer densities (Fig. 7.21). Further variation in winter numbers is caused by winter breeding when acorn crops are very heavy but this does not happen very often (Smyth, 1965).

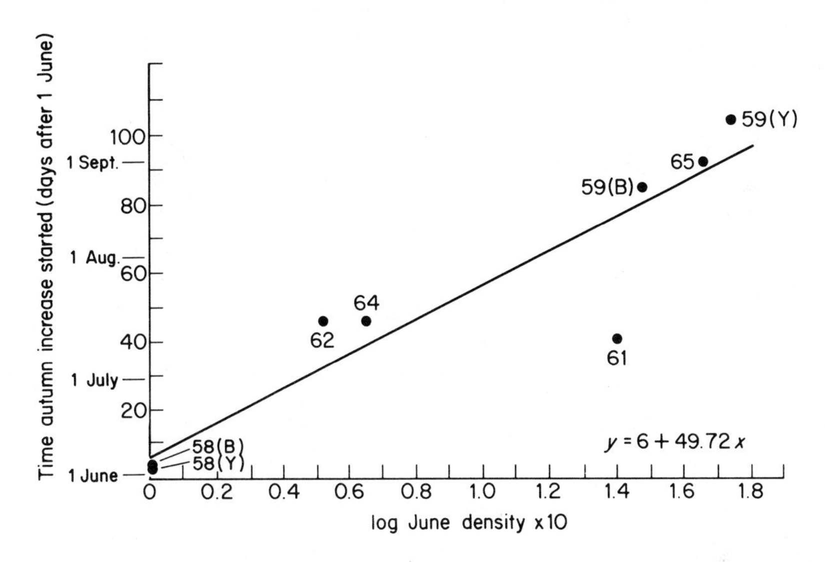

Fig. 7.21 The relationship between the time when the autumn increase in wood mouse numbers started and the density of wood mice at the start of the breeding season. Figures indicate years. Note that B and Y are different study areas. Redrawn from Watts (1969).

The density of the breeding population in spring, having been determined by the overwinter food supply, is thought to be kept constant by the agonistic behaviour of adult males preventing the recruitment of juveniles. When adult males were removed from the breeding population (Flowerdew, 1974) juveniles and subadults survived better and in the first of the two years of the removal experiment numbers increased to the winter peak immediately rather than remaining relatively constant and increasing later as on the control area (Fig. 7.22). These results support the idea that adult male interactions with juveniles are responsible for the poor survival of juveniles during the summer. Further corroboration of this conclusion comes from studies of the behaviour of adult males (Gurnell, 1978) which show that the aggressive behaviour between males which is associated with a home cage in the laboratory increases at the start of the breeding season; also, in the laboratory adult males are detrimental to the survival of juveniles and in the field artificial immigrants survive better in areas where adult males have been removed than in control areas (Flowerdew, 1974). Similar evidence has been obtained to explain the poor juvenile survival of deer mice (*Peromyscus maniculatus*) in North America (Sadleir, 1965; Healey, 1969; Fairbairn, 1977b). By the end of the breeding season the overwintered adult wood mice which were aggressive towards juveniles have mostly died and so the young survive well into the winter, often at the same time as the increasing food supply from the acorns.

Thus social behaviour is implicated in the maintenance of a stable summer breeding population density in the wood mouse and the relaxation

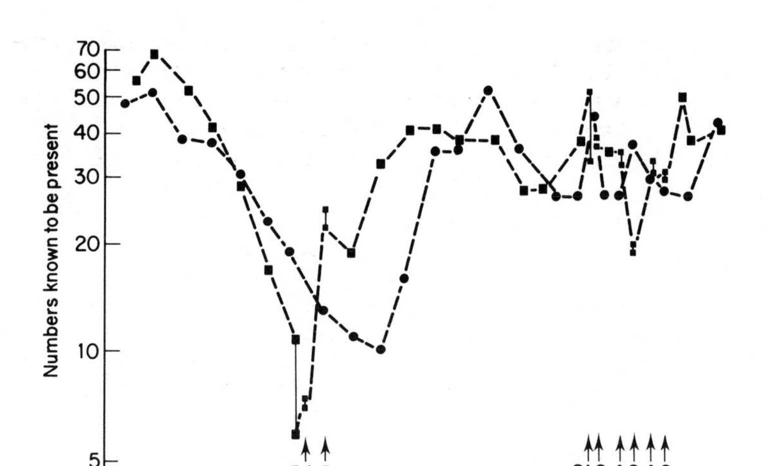

Fig. 7.22 Numbers of wood mice known to be present on a control area (•) and adult male removal area (▪) in woodland near Oxford, between October 1967 and November 1969. Only the 5-week trappings are shown, except at times of special interest. The numbers of adult males (> 21 g) removed (———) are given below the points. Redrawn from Flowerdew (1974).

of this social inhibition of population increase allows winter numbers to be normally much higher than those present in the breeding season. Studies such as these on small mammals lack the detail of observational studies but the pieces of the jig-saw are starting to fall into place. A study of the movements of wood mice using radio-tracking (Wolton, 1985) has shown that the females maintain mutually exclusive home ranges and are probably territorial but that the ranges of adult males overlap greatly. Thus females as well as males may affect juvenile survival, the former by territoriality and competition for space and the latter by hierarchical behaviour and competition for females.

Social behaviour and spacing behaviour are thus seen to be potent forces for limiting population increase and structuring the population so that some individuals remain and some die or disperse.

Dispersal

The importance of dispersal, (movement away from a home area to settle in a new area) to population demography has come to the attention of mammal ecologists only in the last 10–20 years. During this time the biology of dispersal, particularly in the smaller mammal species, has attracted much attention (Lidicker, 1975; Gaines and MacClenaghan, 1980; Stenseth, 1983). The operational definition of a disperser, as an individual entering a vacant area where initially all residents have been removed, has attracted some criticism. This stems from the possibility that short movements outside the home range cannot be distinguished from dispersal movements and this should always be born in mind in the interpretation of such studies. Alternatively, a study of dispersal may allow the identification of dispersing immigrants and emigrants (e.g. Dobson, 1979 working on ground squirrels, *Spermophilus*).

Dispersal allows selective and numerical losses from the population so that its functions may include the avoidance of inbreeding and the avoidance of competition for mates as well as the possible regulation of population numbers (Dobson, 1979). The demographic effect of dispersal will depend on the timing and numerical magnitude of the individual movements; its fundamental importance to small mammal studies was demonstrated when dispersal was 'frustrated' (see p. 173) in enclosed populations of voles (*Microtus* spp.) (Krebs *et al.*, 1969; Boonstra and Krebs, 1977), so that numbers increased markedly in comparison with control, unfenced, areas, and the voles overate their food supply and declined sharply (Fig. 7.23). This has become known as 'the fence effect' and it demonstrates that vole populations cannot regulate their numbers below a food limit if dispersal is prevented.

Studies of dispersal often show that the rate of dispersal increases with the rate of increase of the population as in the Townsend vole (*Microtus townsendii*) and the meadow vole (*Microtus pennsylvanicus*) (Krebs *et al.*, 1976; Gaines *et al.*, 1979). However, this has the consequence that declining populations do not show much dispersal and during this period of cyclic or annual fluctuations (see p. 183) the mortality must be mainly from deaths *in*

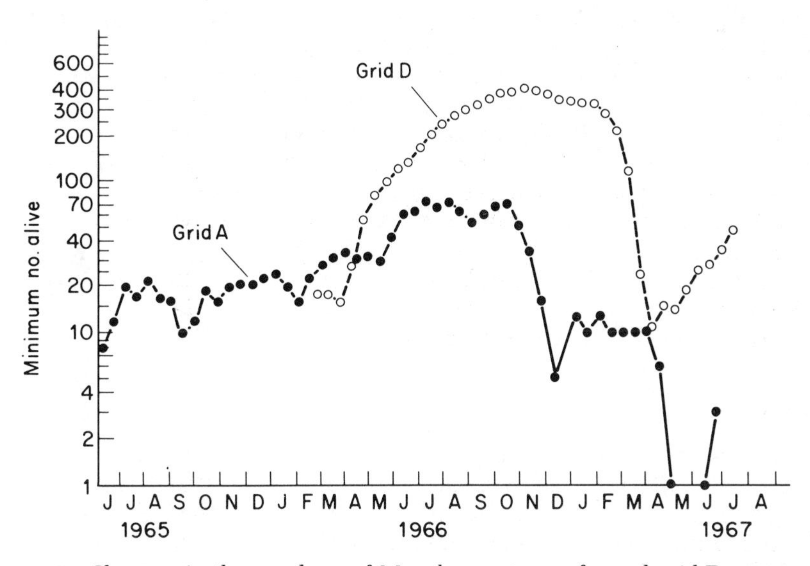

Fig. 7.23 Changes in the numbers of *M. ochrogaster* on fenced grid D contrasted with those on unfenced grid A, 1965–67. Eighteen voles were introduced into vacant grid D on February 22, 1966. Redrawn from Krebs *et al.* (1969).

situ. This may not necessarily apply to both sexes; during the spring decline female wood mice (*Apodemus sylvaticus*) appear to die *in situ*, whereas many males disperse at this time (W. I. Montgomery, personal communication). Similar results have been obtained for the deer mouse (*Peromyscus maniculatus*) in North America (Fairbairn, 1978a,b), where light-weight non-breeding males dispersed in the spring and summer, juveniles and breeding males dispersed at the end of the breeding season, and light-weight mice of both sexes dispersed over the winter.

Why individuals disperse is debatable; presumably it is adaptive and occurs because of social pressure from the resident individuals or perhaps in response to the local limitation of some resource. In a study of the snowshoe hare (*Lepus americanus*) in Canada dispersal of the younger individuals took place at all times, and at high density when there was some evidence of food shortage; light-weight individuals with evidence of scarring (thought to be the result of wounding in intraspecific encounters) moved into the removal areas. It is suggested that these dispersing individuals had a higher mortality rate than residents (Windberg and Keith, 1976).

The study of dispersal has developed hand-in-hand with the study of vole population ecology, particularly that concerned with cyclic fluctuations in numbers (see p. 175). Lidicker (1962, 1975) suggests the types of dispersal and the reasons for them; the types are summarized in Fig. 7.24. He proposed the term 'saturation dispersal' for the outward movement of surplus individuals from a population at or near its carrying capacity. Such individuals are faced with the choice of staying in the population and almost certainly dying, or of moving out and probably dying; they would be social outcasts, juveniles, and very old individuals, in poor condition and least able to cope with local conditions. They would then have only a small chance of reproducing successfully. 'Pre-saturation dispersal' is movement before the habitat is saturated with the species; it occurs during population growth and may occur as soon as growth starts. The dispersing individuals moving at this time will not be in poor condition and may be of any sex or age group, including pregnant females. They will have a much better chance of surviving and breeding elsewhere than saturation emigrants and the advantages of dispersing at this time would be to increase the opportunities for mating,

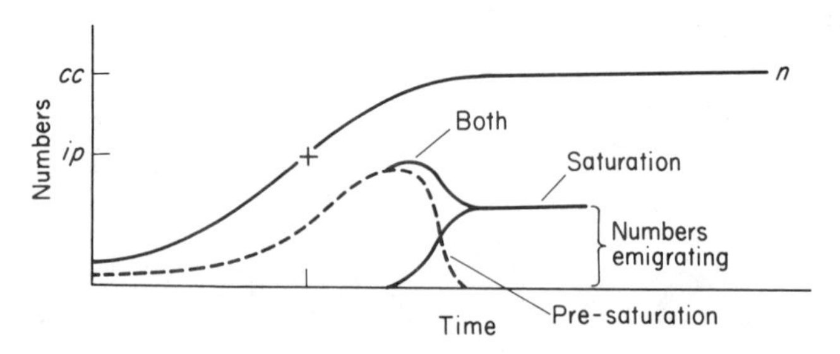

Fig. 7.24 Relation between the changes in numbers (n) of a population and both saturation and pre-saturation types of emigration (dispersal); cc is carrying capacity and ip is inflection point. Redrawn from Lidicker (1975).

to avoid inbreeding, to avoid population crashes and the build-up of predators, and to avoid the inefficient utilization of the remaining resources at high density. Generally, pre-saturation dispersal will occur and few examples of saturation dispersal are available because of the difficulty of defining a saturated habitat.

The role of dispersal in population regulation will, according to Lidicker (1975), vary between species. One example would be when emigration contributes to regulation by saturation dispersal, but dispersal is not essential for regulation as some other form of mortality would soon increase and compensate for the lack of emigration; this is assumed to happen in the Arctic ground squirrel study (Carl, 1971) discussed earlier in this chapter (p. 166). In other species emigration (saturation dispersal) may be important in preventing population growth beyond the limit of carrying capacity; this would need an effective 'dispersal sink' (empty or unfilled habitat) to accommodate long-term emigration. Lidicker suggests that this happened in a population of feral house mice (*Mus musculus*) (Pearson, 1963). In this case if dispersal was frustrated numbers would increase further and then crash. A third example, in practice difficult to distinguish from the second, would be when emigration prevents numbers from reaching carrying capacity by pre-saturation dispersal as probably happens in many *Microtus* species such as those displaying the fence effect (see Fig. 7.23), and in prairie dogs (*Cynomys*) where yearling males and older females emigrate to large unoccupied areas from the breeding colonies (King, 1955).

The terminology used by Lidicker has been criticized by Stenseth (1983) because of the difficulty in obtaining precise definitions; he suggests that 'adaptive dispersal' is pre-saturation dispersal plus some of the dispersal taking place at high density (saturation dispersal); this type of dispersal is independent of density and adults or young emigrating would be healthy and the heritability of this type of movement would be high. Non-adaptive dispersal would be shown by individuals forced out of their home population by competition from superior animals; they are the surplus individuals and include many in the saturation dispersal category, being non-reproductive, of all ages, and having no heritability of dispersal tendency. The differences in the two approaches seem to depend on views of the adaptive nature of dispersal and are difficult to test; whether Stenseth's views are taken up by population ecologists remains to be seen.

Cyclical changes in numbers

Cycles of abundance occur in microtine rodents and have been studied particularly in lemmings (*Dicrostonyx* and *Lemmus* spp.) and in voles (*Microtus* spp. and *Clethrionomys* spp.). Also, an 8–11 year cycle occurs in the North American varying or snowshoe hare (*Lepus americanus*) over much of its range and in a number of its mammalian and avian predators (see Fig. 7.13).

The history of rodent 'plagues' is documented by Elton (1942) and the search for the cause(s) of the cyclic fluctations has produced an enormous literature (see reviews by Krebs and Myers, 1974; Tapper, 1976; and

Finerty, 1980). It is useful to concentrate here on the microtine cycle where most experimental work has been carried out to test the various hypotheses put forward in explanation.

In the microtine cycle the fundamental questions to be answered are 'what causes the 3–4 year periodicity?' and 'how is synchrony brought about over wide geographical areas?'. However, the same species may be cyclic in one geographical location but not in another, for example the bank vole (*Clethrionomys glareolus*) is cyclic in northern Sweden but in southern areas it shows only annual fluctuations in numbers (see review by Alibhai and Gipps, 1985); thus any explanation of cyclic fluctuations must also explain the shorter fluctuations which are commonly observed. Even after thirty years of research it is evident that only part of an answer can be provided to the questions although some hypotheses can now be more firmly rejected than in the past.

The characteristics of cyclic fluctuations (Fig. 7.25) as outlined by Krebs and Myers (1974) can be seen in four main phases.

(1) The increase phase which shows a large increase in numbers from very low density in one spring to the next, and may last for 2–3 years as in many brown lemming (*Lemmus trimucronatus*) populations, or may be a

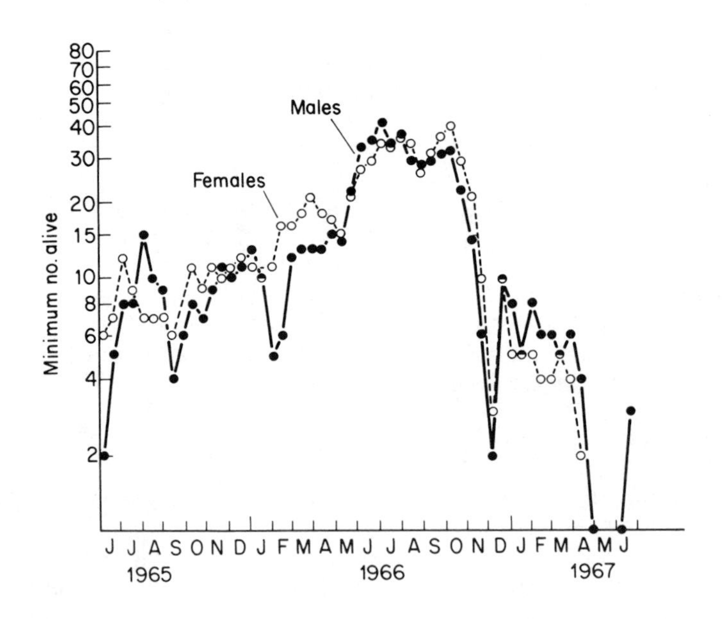

Fig. 7.25 An example of one multiannual 'cyclic' fluctuation in population numbers. Population changes in *Microtus ochrogaster* on an unfenced grid, 1965–67. Redrawn from Krebs *et al.* (1969).

rapid 'explosion' of numbers in one year as found in many *Microtus* species in Europe and North America.

(2) The peak phase, which shows little change in numbers from one spring to the next, or as in some *Microtus* species lasting only a few months.

(3) The decline phase, which may last for up to two years and may show a temporary recovery in the breeding season, or alternatively may occur rapidly over one year with no recovery or 'crash' from winter to spring after a peak year.

(4) The phase of complete scarcity where, if it occurs at all, populations fall to low numbers for up to three years.

During the cycle the litter size tends not to vary significantly and the pregnancy rate does not change; however, the length of the breeding season and the age at sexual maturity may change. The increase phase of many vole and lemming populations occurs during a period of winter breeding or during an extended breeding season. During the peak phase individuals mature at older ages and heavier weights; during the decline phase maturity may be delayed. Mortality in adults is low in the increase and peak phases but higher in the decline; however, an increase of as little as 10–15% in mortality can cause a decline. Juvenile mortality is high in the peak and the decline. Dispersal appears to be greatest during the increase phase and smallest in the decline phase, at least in *M. pennsylvanicus* (Myers and Krebs, 1971).

It has recently become a subject of debate whether immigration is needed as well as emigration to allow cycles to occur (Abramsky and Tracy, 1979; Rosenzweig and Abramsky, 1980). A consequence of immigration is suggested to be the alteration of the amount of genetic heterozygosity and the associated amount of aggressive behaviour (populations with greater heterozygosity are shown to be more aggressive; Smith, *et al.*, 1975), or alternatively the amount of genetic variability (Abramsky and Tracy, 1979), without which cycles are supposed not to occur. However, Tamarin *et al.* (1984), using an enclosed population which included a dispersal 'sink' of wooded habitat, have shown that immigration is not necessary to produce one cycle of density in *M. pennsylvanicus*, although they admit that immigration may be necessary for further cycles to occur.

Hypotheses to explain cyclic fluctuations in numbers may be divided into four main categories dealing with food, predation, behaviour and 'multifactorial' ideas. None of these provides a satisfactory answer but many aspects of microtine population dynamics are now more fully understood.

Food quality and quantity have both been suggested as the cause of microtine cycles. Food quantity was proposed by Lack (1954) as a cause of lemming cyclicity; overexploitation of the habitat would result in the destruction of cover and a greater loss by predation than before. Alternatively, Pitelka (1957) proposed that the lack of food produced by high lemming density led to malnutrition and reduced reproduction and so to a decline. These food quantity hypotheses are refuted by evidence from (a) lemming (*Dicrostonyx*) populations at high density where the food supply is not overeaten (Krebs, 1964), (b) by the fence effect in *Microtus* species (see p. 171), where overexploitation of the food supply is shown to be abnormal, and (c) experiments where additional food has not prevented a decline in population numbers, for example in *M. californicus* (Krebs and Delong, 1965), in *M. ochrogaster* (Cole and Batzli, 1978; see Fig. 7.9) and in *M. townsendii* (Taitt and Krebs, 1981).

Food quality was introduced in the 'nutrient recovery' hypotheses put forward by Kalela (1962), Pitelka and Schultz (1964) and Tast and Kalela (1971); these suggest that it takes several years to recover from previous grazing or flowering. The hypotheses were extended by Laine and Henttonen (1983) who report a correlation between the increase phase of the populations of five microtines in northern Fennoscandia and peak plant production (thus flowering, berry and seed production is cyclic). Good feeding conditions in the increase phase are followed by a reduction in the nutritive and energy value of the food plants in the second winter after the flowering peak; this occurs as a result of overgrazing and the induction of plant defence secondary compounds which make them unpalatable. Extrinsic factors such as predation by small mustelids and disease cause the decline, making the hypothesis multifactorial (see p. 136) and it is still untested.

The nutrient quality hypothesis of Pitelka and Schultz (1964) and Schultz (1964) was suggested to limit lemming population growth. In the summer of peak numbers the nutrients, particularly phosphorus and calcium in the food are at a high concentration and the high production and high consumption by the lemming population 'locks' the nutrients into organic material so that they are unavailable to plant growth in the following year. Thus after the peak year nutrient levels in the food are low and this inhibits reproduction in the lemmings. By the fourth year of the cycle enough nutrients have been released to allow the cycle to enter the increase phase again. Krebs and Myers (1974) criticized this theory because phosphorus levels in the forage were similar before and after the peak density. They also found no evidence of phosphorus deficiency limiting reproduction or of calcium levels limiting lactation.

Toxic compounds in food plants were also suggested by Freeland (1974) to be the cause of cyclic mortality. At peak densities mortality could be caused by reduced viability induced by toxic compounds; these chemicals may inhibit protein synthesis, growth and/or reproduction. Thus in a cyclic population of voles the following would occur: (a) voles would prefer non-toxic foods, (b) as population density increased there would be a reduced availability of non-toxic foods and an increased availability of non-preferred toxic foods, (c) at high density there would be increased ingestion of toxic foods with a subsequent reproductive decline and reduction in population density, and (d) with the population decline there would be reduced cropping and preferred foods would outcompete toxic foods so that another increase phase could start. However, Batzli and Pitelka (1975) concluded that there is no clear idea of what is toxic to voles or lemmings and that the same species at two sites with different vegetation both appeared to cycle in phase, so casting doubt on Freeland's ideas. Further detailed criticisms have been made by Schlesinger (1976) but the results of Myllimaki (1977) and Bergeron (1980) lend support to this hypothesis. The specific plant compounds which influence microtine reproduction are discussed in Chapter 3. It must be concluded from the above that there is little convincing evidence to show that food quality or quantity can cause cyclic fluctuations.

Predation as a factor in mammalian population dynamics has already been discussed in an earlier section of this chapter. In cyclic fluctuations of rodents Pearson (1971) suggested the following after a study of predation on *Microtus californicus*.

(1) Avian predators are not intensive enough in their predation to determine population trends in their prey since they leave the area when prey are at low density.
(2) Predators are unable to prevent the increase of breeding microtine populations.
(3) The amplitude of the cycle may be increased by predators since predator pressure continues at very low prey population densities, forcing numbers even lower.

In reviewing the predation hypothesis, Myers and Krebs (1974) emphasize that some authors claim that predators take a high proportion of the prey population and are able to prevent population increases, while others suggest that predators have very little influence on the prey population; in both cases comprehensive quantitative studies are lacking. The point is also made that two species of microtine often cycle in phase but the decline of one species or sex may well precede that of the other by several months. If predators are causing the mortality during the decline they must be highly selective; this could be the case but it seems unlikely that the selection can be so one-sided (see section on predation). Experimental studies of predation (Taitt and Krebs, 1983) indicate that declines can occur without predation but the role of predators in the low phase is still to be elucidated.

The influence of predation on microtine populations is likely to be variable, according to the type of predation occurring. It was suggested earlier (see p. 154 and Erlinge *et al.*, 1983) that the non-cyclic populations of *Microtus* in southern Scandinavia fluctuated only seasonally because many avian and mammalian predators were facultative and were sustained during population lows by alternative prey; these predators showed a strong functional response to the increase in prey population density and so did not allow cyclic fluctuations. In northern Scandinavia the prey populations cycle because the smaller number of predators are obligate, owing to the smaller number of alternative prey, and so the predators follow the prey cycle with a strong numerical response which will probably be delayed. A similar explanation (see p. 184) is put forward for the lack of cyclicity in showshoe hare populations in the southern parts of its range (Wolff, 1980). These are convincing arguments concerning the role of predators in dampening the amplitude of microtine population fluctuations but they are the result of descriptive studies and in themselves they do not prove that predation drives the cycle.

Behavioural hypotheses involve spacing and aggressive behaviour and may be divided into phenotypic and genotypic types. All the behaviour hypotheses are associated with the criteria drawn up by Watson and Moss (1970) (see p. 162) and depend on the existence of surplus individuals and the possibility of dispersal, although this has not always been admitted. Changes in the spacing behaviour of individuals are necessary to prevent the unlimited increase of the population and so the breeding population is limited.

The 'Stress' hypothesis has become a widely quoted example of the phenotypic behaviour type and has gained continued support since the 1950s (Calhoun, 1949; Christian, 1950, 1971, 1980; Christian and Davis, 1964). The stress or neurobehavioural-endocrine hypothesis states that mutual interactions lead to physiological changes (phenotypic in origin) which reduce births and increase deaths, and so the population declines (Krebs, 1978a). This hypothesis was derived from the work of Selye (1946) who proposed the General Adaptation Syndrome, an integrated syndrome of adaptive reactions to stress, involving physiological and morphological changes and ending in exhaustion and low resistance to disease and other mortality factors if the stress is maintained. Changes in the adrenal glands and the pituitary gland are involved in the endocrine reactions to the stress.

Christian's laboratory observations led him to believe that increased social interaction at high density leads to physiological changes including the hypertrophy of the adrenal glands and the increased activity of the adrenal cortex which produces glucocorticoids. Hypertrophy of the spleen is also induced by stress, possibly as a reaction to increased adreno-corticotrophic hormone (ACTH) from the pituitary or as a reaction to pathogens; somatic growth is suppressed and reproduction curtailed; involution of the thymus (as a response to corticosteroids) may lead to reduced vigour and reduce the immunological response. Thus, as Krebs and Myers (1974) suggest, a test of the stress hypothesis would be to find evidence of increased adrenal activity at high density and individuals in poor physiological condition during the decline. Initially, adrenal weight was used as an index of adrenal activity (Christian and Davis, 1964), and physiological measures were later developed, but much of the work has been criticized (see Krebs and Myers, 1974). More recently, studies of cyclic, mainland populations of *M. pennsylvanicus* and non-cyclic, island populations of *M. breweri* by To and Tamarin (1977) show that the adrenal weight (mg/g body weight), which is correlated with the size of the glucocorticoid producing zones, did not increase with population density in the cyclic population but did show an increase correlated with the increase in population size in the non-cyclic, high density, island population. This is interpreted as showing that the stress hypothesis is not responsible for limiting population growth in *M. pennsylvanicus*. However, contrary to these results, Geller and Christian (1982) found that the adrenal weight of mature female *M. pennsylvanicus* was correlated with the mean population density in spring (April–June). Further tests of the hypothesis are needed.

There seems no doubt that stress reactions may occur in rodent populations but what is controversial is their relevance to normal population regulation in cyclic and non-cyclic species. Thus brown rats (*Rattus norvegicus*) in a land-fill area showed classic stress symptoms (Andrews *et al.*,

1972); however, Lidicker (1978) asserts that although stress is likely to be a potential physiological mechanism for mammalian population regulation, it must also react with environment variables such as space, and individuals in the field would not normally be expected to show the stress syndrome. Animals which become unduly stressed before resources run out would be selected against. Stress is selected for to meet normal physiological changes, but the occurrence of stress as an emergency measure when population density is high should not have evolved as a population regulating mechanism. Physiological changes may also occur in natural populations by other mechanisms, for instance as a result of olfactory cues (p. 81), but the possibility of dispersal may allow such changes to be reversed.

A case of stress being part of the natural annual cycle of population changes is seen in the Australian marsupial mouse (*Antechinus stuarti*). Lee *et al.* (1977) showed that all males in the population die within three weeks of the start of the August mating period. Males caught earlier and caged singly live well beyond the time of natural mortality. Studies of the free and albumin-bound corticosteroid levels of the males showed that they were higher in August than in July and also higher than those of females in August; it is suggested that the population-wide death of males following mating results from too much corticosteroid suppressing the immune system and inducing death from disease. Adrenal hypertrophy is caused by the aggressive behaviour of the males and it has also been found that the binding sites for corticosteroids in males are fewer than in females. Thus stress plays a part in natural population changes, but the burden of proof for the relevance of the stress hypothesis in cyclic fluctuations is on Christian's supporters.

Another phenotypic behaviour hypothesis is Hestbeck's (1982) 'social fence' hypothesis. This suggests that when population density is low individuals survive in areas of good habitat and as population numbers increase aggressive interactions between individuals increase (Turner and Iverson, 1973) and those which disperse remain in neighbouring, previously uninhabited areas. Population density then increases in refuge and neighbouring areas, and at higher population density the aggression encountered from neighbouring social groups is greater than that within social groups so that dispersal is effectively inhibited; population regulation can then no longer be achieved by dispersal and must operate by resource exhaustion, the social groups being 'fenced in' by neighbouring social groups. In areas of heterogeneous habitat, populations are regulated alternately by spacing behaviour and resource exhaustion, and this is postulated to lead to cyclical fluctuations; the difficulty with the idea is that resource exhaustion leads to the decline and evidence for this is poor (see above). Although this is mainly a behavioural hypothesis it approaches a multifactorial (see below) hypothesis because it integrates behavioural and resource exhaustion as causes of cycles.

An hypothesis which bridges the phenotypic and genotypic behavioural hypothesis is Charnov and Finerty's (1980) 'kin-selection'. This relates aggressive behaviour and kin-selection so that the aggressive behaviour of individuals is all that is needed at high density to precipitate a decline. At low density there is little dispersal and most individuals will be amicable to their neighbours because they are likely to be relatively closely related to each other (high coefficient of relationship); at high density many less well related individuals interact and there is more aggressive behaviour, leading to a decline. This hypothesis does not involve a change in the genetic composition of the population, just a reduction in the relatedness of individuals at high density, so it is not a true 'genetic' hypothesis. Indeed, critics of the hypothesis (Bekoff, 1981) suggest that it is simply familiarity, not the genetic coefficient of relatedness, which is important in mediating the nature of the social interactions; familiar individuals show less fighting and display more cohesive, amicable, behaviour than less familiar individuals at high density: recognition is learned rather than inherited.

The true genetic behavioural hypothesis, also known as the polymorphic behaviour or Chitty's hypothesis, attempts to explain how any population may be limited and so applies to cyclic and non-cyclic populations. It was first proposed by Chitty in the 1950s and refined later (1960, 1967). The mechanism of self-regulation was proposed to be a consequence of natural selection under conditions of mutual interference, so that some genotypes have a worse effect on their neighbours than vice versa (Krebs, 1978a; Krebs and Myers, 1974). Spacing behaviour is the mechanism which limits population density as in the phenotypic behaviour hypotheses, but the change in spacing behaviour has a genetic basis so that the behaviour responds to natural selection in a short period of time. The specific predictions of the hypothesis and tests of them are reviewed by Krebs (1978a); a version of the hypothesis, as modified by Krebs and Myers (1974) is presented in Fig. 7.27. As the population increases aggressive genotypes are selected for because they have an advantage at high density, but they tend to show less successful reproduction and a decline is the result. During the decline less aggressive genotypes are selected for which are more tolerant of others and these have a high reproductive potential so that the population will start to increase again. Thus spacing behaviour is the driving force behind the cycle, but changes in behaviour are the result of genetic selection which also changes through the cycle. The social intolerance of increasing populations leads to a high rate of dispersal as well as reduced fecundity and an increase in the death rate; aggressive individuals dominate the population at the peak but they prevent the population from increasing further. Only when these aggressive individuals have decreased in numbers and aggression has ceased to be an advantage will numbers start to rise again.

The many tests trying to substantiate Chitty's hypothesis have been only partly successful. Early studies (e.g. Tamarin and Krebs, 1969; Gaines and Krebs, 1971) showed that 'genetic markers', such as blood serum proteins, were polymorphic and that the frequency of each morph changed with the cycle. In the case of transferrins in *M. ochrogaster* the homozygotic polymorph (EE) appeared to be advantageous in the increase and the heterozygote (EF) was usually advantageous in the decline (homozygote FF was rare). At the same time studies of behaviour (Krebs, 1970) showed that aggression changed with population density in both *M. ochrogaster* and *M. pennsylvanicus*; male *M. ochrogaster* from peak populations were most

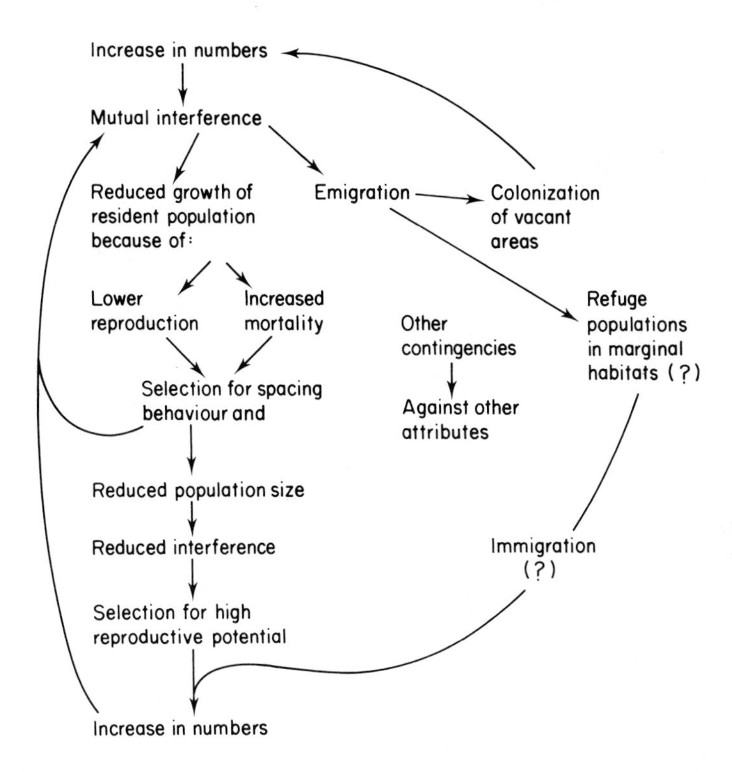

Fig. 7.26 The Krebs/Myers modified version of the Chitty behavioural-genetic hypothesis to explain microtine cycles. Dispersal is viewed as being a more important aspect than originally proposed by Chitty. Selection acts through behavioural interactions and the changing genetic composition of the population with fluctuating densities. Redrawn from Krebs and Myers (1974).

aggressive and home range size was related to aggressive behaviour. However, the behavioural attributes were not linked to the transferrin genotypes of either species. In another study of the same species (Myers and Krebs, 1971), it was shown that dispersing males were more aggressive and less active than resident males, and that frequencies of polymorphic plasma proteins (transferrin – Tf, and leucine amino peptidase – LAP) differed in dispersing and resident populations. In addition, young females were more common in the dispersing populations of both species. Thus support was gained for both Lidicker's theories (see Dispersal) of population regulation by dispersal and for Chitty's hypothesis that a behavioural polymorphism regulated population density.

Further tests of Chitty's hypothesis were, however, not as successful; tests of the genetic influence on population demography failed. These experiments manipulated the frequencies of polymorphic blood protein genetic markers detected by electrophoresis (Tf or LAP; see above) in enclosed populations of *Microtus*, so that they were at the same frequencies as in increasing and declining populations (Gaines *et al.*, 1979; LeDuc and Krebs, 1975). No differences in demography were observed between the experimental and control populations; thus when the LAP 'fast' homozygote was maintained at 75% by removing the LAP 'slow' homozygotes and in another area LAP fast homozygotes were removed to give a 25% frequency of LAP fast; these populations and the control (35% LAP fast) all went through the same increase and declined sharply in the spring. Thus the genetic polymorphism studied appeared to be affected by demography but not vice versa. This means that the genetic linkage of blood proteins with behaviour is unlikely and that the genetics of the behavioural attributes themselves should be studied to properly test Chitty's hypothesis. The value of these studies, however, has been to demonstrate the speed of genetic selection in ecological time and that behaviour does indeed change with the cycle and is different in dispersing individuals and resident individuals. At the present time it seems as if spacing behaviour characteristics may not be inherited (Anderson, 1975; quoted by Krebs, 1978a), but the results depend on the relevance of laboratory studies of behaviour to the field situation.

The influence of dispersal and spacing behaviour on population dynamics, whether cyclic or not, has been demonstrated by studies connected with the genetic-behaviour hypothesis. One of these was carried out by Tamarin (1978) during his study of the cyclic mainland *M. pennsylvanicus* and the non-cyclic island *M. breweri* in North America. The study showed that dispersal was necessary for cyclic fluctuations to occur (as in the fence effect control populations; see section on dispersal, p. 170). Although there appeared to be marginal habitat on the island which might act as a dispersal 'sink' (*sensu* Lidicker, 1975) no proper dispersal was taking place because there were no mammalian predators and only low densities of avian predators; this allowed the dispersing animals to return to the resident population, and numbers remained at high density rather than show the expected cyclic fluctuations, as the other species did on the mainland, where dispersal is effective and dispersing individuals are presumably preyed upon by mammalian predators. Tamarin goes on to suggest that dispersal, predation, and natural selection should be integrated, by placing them all on the same continuum. Dispersal is suggested to be increased by predation and so natural selection is intensified; if declines occur on the mainland because dispersal has altered the gene pool, then on the island one would not expect a cycle because the gene pool has not been altered by dispersal. Thus further circumstantial support for Chitty's hypothesis is provided.

Studies of spacing behaviour in the usually non-cyclic *Microtus townsendii* by Krebs (1978b) have shown that surplus individuals (those which move into vacant areas) are present at all times. These surplus individuals show more subordinate behaviour than residents and are probably prevented from breeding by residents. Also, in the same species, further studies by Taitt and Krebs (1983) suggest how high density cyclic peaks may occur as opposed to lower density annual fluctuations. The populations are regulated by food and behaviour (Taitt and Krebs, 1981; see effects of variation in food supply, p. 147), and it is thought that if the voles are nonaggressive prior to breeding and if the food supply suddenly improves in spring then large numbers can become reproductive simultaneously and settlement could be at a relatively high 'pre-cyclic' spring level and lead to a cyclic 'high'; the juveniles produced would be likely also to settle simultaneously at a high density in any temporarily suitable habitat. Normally reproduction starts at the time when food supplies are poor and so only moderate 'non-cyclic' spring densities occur. The essential difference between the non-cyclic and the cyclic fluctuations is in the nature of the spring population; after a substantial spring decline only an annual fluctuation will occur but after a small spring decline a multi-annual cycle occurs. This hypothesis has some support from the work of Birney *et al.*, (1976) who suggest that vegetative cover (and presumably food supply) must be good before densities are high enough to show cyclic fluctuations, and so the ideas tend towards a multifactorial hypothesis.

The multifactorial hypothesis, recently given impetus by Lidicker (1973, 1978) is summarized in the introduction to this chapter. A network of environmental factors interact independently with the population and with each other so that at any one time one or a few factors may be predominant in their effects on density. In his study of the Californian vole (*M.californicus*) Lidicker (1973) concluded that at least six factors – emigration, predation, competition, carrying capacity, physiological damage (delaying the onset of reproduction), and the beginning of the wet season – were involved in explaining the cyclical fluctuations of 3–4 years' duration. At low density empty habitats are filled by pre-saturation dispersal but once they are filled dispersal is frustrated and populations rise to a peak. Any increase is then prevented by a seasonally declining carrying capacity and numbers are reduced by the exhaustion of resources, lowered reproductive rates, predation and saturation dispersal. This type of hypothesis has been criticized (see p. 134) because it could only be substantiated by the association of events over a long period of time (Krebs, 1978a); however, some authors claim it has led to a breakthrough in the thinking about cyclic fluctuations in that research workers are now prepared to accept the possibility of more than one factor operating to cause cyclic fluctuations. Further developments are awaited with interest!

The 10-year cycles observed in many showshoe hare (*Lepus americanus*) populations may also have a multifactorial cause (see similarities with microtine cycles discussed by Keith, 1974 and Lidicker, 1978). In reviewing the role of food in hare cycles, Keith (1983) (see also sections in this chapter on predation p. 148 and food supply p. 137) suggests that a predictable syndrome of birth and survival rate changes resulting from overwinter food shortage initiates a cyclic decline of the snowshoe hare in North America and probably also of the Arctic hare in the Soviet Union. Predators become the dominant mortality factor in snowshoe hares only after the initial food-related declines have greatly increased the ratio of predators to hares and juvenile survival (and reproductive rate) remains low for 2–3 years, recovering over a further 2–3 years. The increase is caused by a rise in juvenile and adult survival after reproduction has recovered. The inter-regional synchrony is caused by mild winters which moderate mortality in peak populations and permit lagging populations to attain peak densities; this is reinforced by the highly mobile predator populations concentrating on the highest densities of hares. Both the hare–vegetation and the predator–hare interactions have a strongly delayed density-dependent nature which promotes and sustains the cyclic oscillations.

The overwinter (6–8 month) diet of snowshoe hares is mainly twigs of shrubs and small trees. At peak densities there is insufficient food to support the hare population and this shortage will remain at least through the next winter; this has been shown in Alberta and in Alaska and by experimental manipulation of the food supply. The food shortage may be prolonged by the browsing-induced production of terpene and phenolic resins which repel hares and so reduce the food supply during the decline after severe browsing at the peak. It is proposed that non-cyclic hare populations occur where suitable habitats are fragmented, such as in the montane ranges and at the southern limit of the hare's distribution: in such areas the amplitude of the cycle is reduced because dispersers from the islands of favourable habitat are removed by many facultative predators (cf. Erlinge's ideas concerning microtine fluctuations discussed above) and so the interaction of habitat, dispersal and predation limits numbers. In contrast, in the large areas of favourable habitat, with food and cover, dispersers survive better and as the predators are obligate they tend to have a delayed cycle dependent on the hares (see Fig. 7.13). Whether, in fact, the density of non-cyclic peaks is less and food shortage non-existent in noncyclic fluctuations is debatable.

Conclusion

It is obvious from the foregoing sections in this chapter that there are still many unsolved problems in mammalian population biology, even in the few species that have been studied intensively. The search for a cause of cyclic fluctuations will undoubtedly continue and on the way mammalian population ecology will benefit, whether a convincing answer is found or not.

8

Man and Mammalian Populations

This chapter introduces some of the ways in which man interacts with natural mammal populations. Particular attention is paid to population simulation techniques under different circumstances, including both exploitation and conservation, and to how they may be useful in quantifying the nature of the interaction for the benefit of both man and mammals.

Models for the exploitation of whale and seal populations

Whaling has a long history, most of which is a tribute to man's greed for financial gain (Bonner, 1980; Gambell, 1976a), and is now more a matter of past, rather than present-day, interaction with man, following the recent moratorium on whaling accepted by most whaling nations.

The biological principles which have been used to assess whale stocks were derived from fisheries biology and are examples of ecological theory being used in practice. They have been the subject of criticism but nevertheless were the best approximations available to quantify the situation where unrestricted exploitation could, and has, severely depleted the stocks of whales all over the world.

Central to the management of whale stocks was the assessment of population numbers for the various species. This proved to be a difficult task, as although the catch records were good the effort used in the exercise was difficult to estimate accurately and many other factors seemed to confound the various methods available for population estimation (Gambell, 1976a). The International Whaling Commission, the body set up to advise the whaling nations on acceptable catch quotas, had to give advice based on accepted estimates of numbers, recruitment and mortality, and errors made here may have had unfortunate consequences for the whale populations (May, 1976). The advice was aimed at setting quotas for the numbers of whales which could be removed without depleting the populations.

The model used to estimate the catch quotas was based on the 'sustainable yield curve' which is developed as follows (Gambell, 1976a,b). It is assumed that a population of whales is a single homogeneous unit at equilibrium at the carrying capacity of the environment, allowing that minor fluctuations may occur. Under these circumstances the recruitment of new individuals to the population should be equal to the natural mortality, and the numbers in the population would remain constant. When some animals are removed (by whaling) the population number is reduced and it will

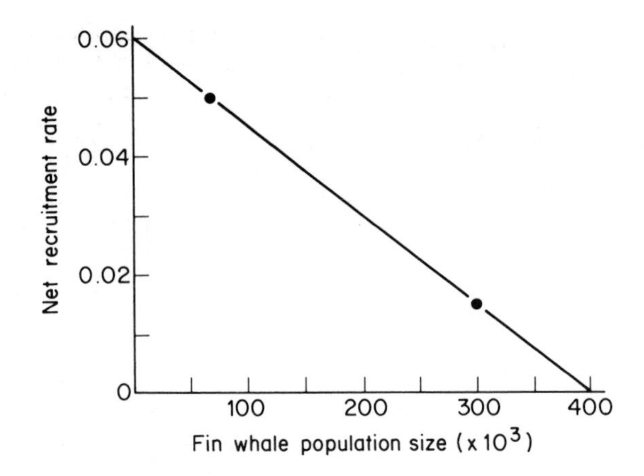

Fig. 8.1 Possible relationship between net recruitment rate and the Antarctic Fin whale population size. Redrawn after Gambell (1976a).

increase at a rate dependent on the abundance (density) of the population and changes in the recruitment and mortality rates. Thus additional recruits would be added to the population to bring it back to the equilibrium level, but if these additional recruits were removed again by whaling then the population would remain at a reduced level and continually produce additional individuals at the rate described above; thus a catch could be sustained indefinitely. This net recruitment rate will decrease as the population size increases (Fig. 8.1), and when this rate is multiplied by the population size to which it applies, it gives the 'sustainable yield' for the appropriate population size (Fig. 8.2). Note that the yield is small for both small and large population sizes because of the small population size in the former case and because of the density-dependent reduction in recruitment

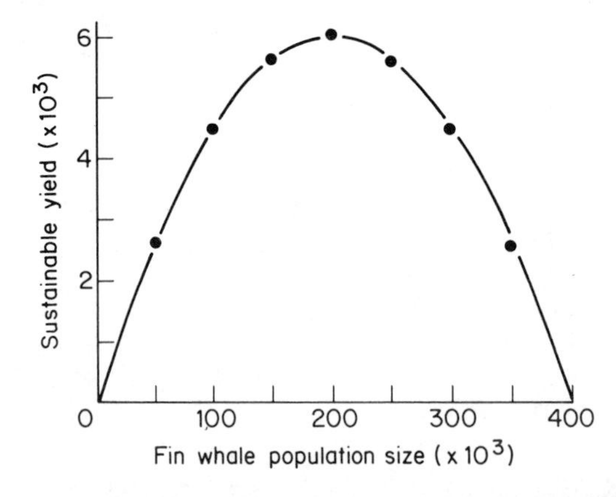

Fig. 8.2 Possible sustainable yield curve for Antarctic Fin whales derived from the recruitment relationship (Fig. 8.1). Redrawn after Gambell (1976a).

rate in the latter. At an intermediate level of population size the sustained yield reaches a maximum (maximum sustained yield, MSY), and this would be the level aimed for in successful continued commercial exploitation of the population, as it causes no change in the population and allows maximum catches; this level is often assumed to be at 50% of the maximum (equilibrium) numbers in the population. At population levels lower than the MSY level a catch lower than the sustainable yield will allow population numbers to increase and if the catch is larger than the sustainable yield numbers will decrease. At population levels larger than the MSY level, catches lower than the sustainable yield will allow the population to increase further but have a lower sustainable yield, and if catches larger than the sustainable yield are taken then the population will support this level of exploitation as long as it is lower than the MSY level. On the right-hand side of the MSY level as the population decreases the sustainable yield increases until the MSY level is reached and if MSY catches are maintained the population will continue at this level. The differences between this simple model and the real situation in whale populations (including the probable time lags in the system because recruitment is related to densities

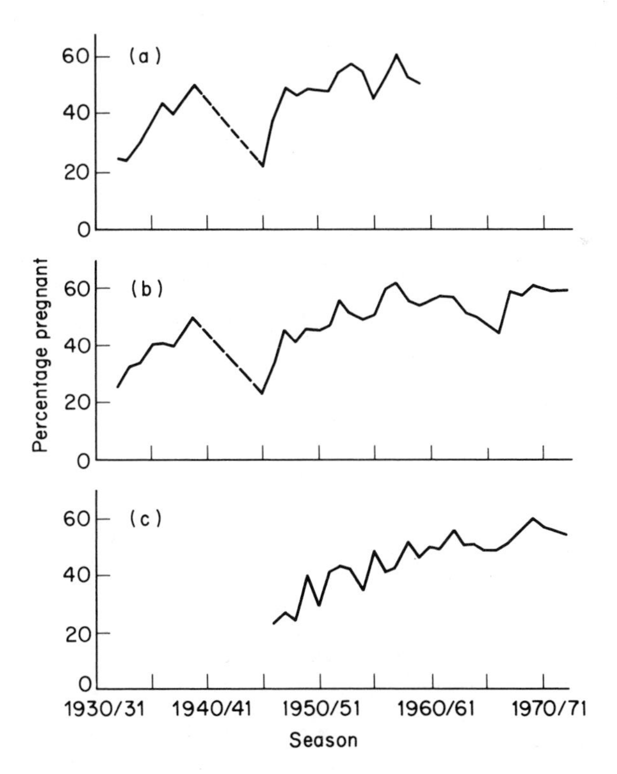

Fig. 8.3 Number of pregnant **(a)** blue, **(b)** fin and **(c)** sei whales as a percentage of the estimated number of mature female whales caught each season in the Antarctic (from International Whaling Statistics, Oslo). Redrawn from Gambell (1976a).

encountered earlier than the time of the current catch) are discussed by many authors (e.g. Allen, 1980; Gambell, 1976a,b; and May, 1976).

In developing the recruitment curve a linear reduction in net recruitment rate with increasing population density is often assumed but this may not necessarily be the case (Gambell, 1976b). It is possible that the reduction or increase in numbers of one species of whale may affect other species by increasing or decreasing their food supply. This is apparently the case when the pregnancy rates of sei (*Balaenoptera borealis*), fin (*B. physalis*), and blue (*B. musculus*) whales in the Antarctic are compared (Fig. 8.3; Gambell 1973; 1976a,b). It seems that the pregnancy rates of the sei whale were increasing before the species started to be exploited in the 1960s; this species shares its krill (crustacean) food supply with the blue and fin whales which had been reduced earlier.

Similar sustained yield models have been developed for the exploitation of the Atlantic harp seal (*Pagophilus groenlandicus*) off the Canadian and adjacent North Atlantic coast, allowing a sustained yield to be estimated for the North Atlantic 'fishing' operation (Lett and Benjamin, 1977; Lett *et al.*, 1979). In one model (Fig. 8.4) the simulation allowed the catch of pups (yield) to vary in response to population size (the basic stock of 1+ (years) age classes), controlling the hunting mortality rate and adding an additional mortality from large offshore vessels of 1%. The hunting level on pups (less

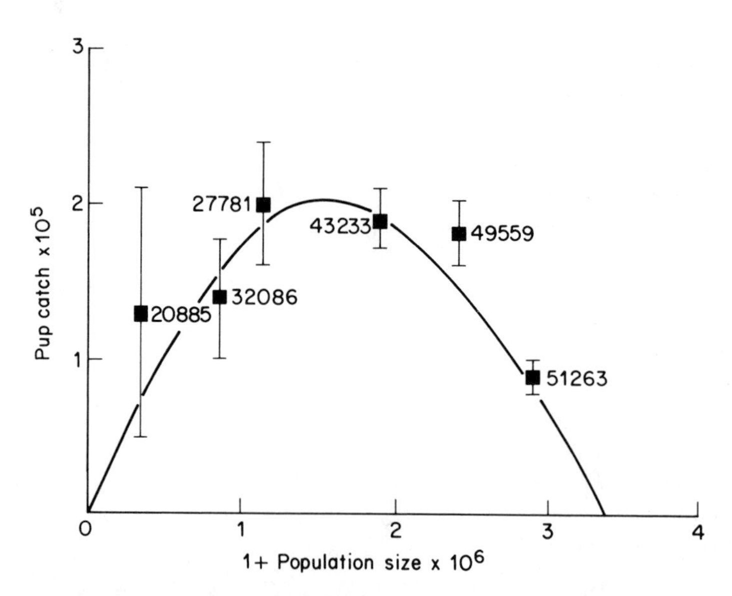

Fig. 8.4 Curve illustrating the MSY population level of ~1.6 million for the catch of pups (squares) and 1 + age group of north-west Atlantic harp seals (figures beside squares) in their 1976 proportions. As the lower catches of pups allow large population sizes, the constant hunting mortality by large vessels will allow a greater catch of 1 + seals. Bars represent 2 × S.D. (standard deviation). Redrawn after Lett and Benjaminsen (1977).

than 1 year old – the pup catch) was allowed to vary and the predicted population level for a MSY was 200 000 pups (and 40 000 1+ seals, given the current killing ratio of pups to 1+ seals); this MSY level was obtained at a total population of 1+ seals of 1.6 million, equivalent to a breeding stock of 375 000 seals. This sustainable yield curve was used in the prediction of changes in the breeding stock with varying management strategies so that the time taken to reach MSY levels of exploitation could be estimated. In addition, the estimated population size in 1977 of 1.2 million was used to set a sustained yield quota or total allowable catch of 170 000 seals for that year.

Theoretically, MSY levels of exploitation should allow continued harvesting of an economic crop from the population and so 'conserve' the population. However, the theory and practice have not always coincided in every population. In the case of the Atlantic harp seals, it appears that numbers have been increasing after a reduction from the levels of the 1950s (Platt, 1976). Arguments have continued over the accuracy of the MSY levels for the harp seals (Winter, 1978; Lett *et al.*, 1979) and later authors are not fully convinced about the density-dependent nature of the reduction in reproduction presumed for the population (Bowen *et al.*, 1981).

The conservation of mammalian populations

Conservation objectives in the management of biological systems are often wide-ranging; the possibilities are outlined by O'Connor (1974) as 'species conservation', a concern for the protection of rare and interesting species, 'habitat conservation', the maintenance of representative habitat types over a range of climatic and edaphic conditions, 'conservation as an attitude to land-use' so that land-use planning and management is organized to minimize the disruption or deterioration of natural systems, and 'creative conservation', the large-scale modification of the landscape by reclamation or development. As far as mammalian populations are concerned, species and habitat conservation have a direct bearing on the survival of some species and on man's deliberate manipulation of populations to prevent extinctions. The other objectives are equally relevant to mammalian populations but are less likely to be carried out with mammalian populations particularly in mind.

Species preservation for mammalian populations usually takes the form of attempts at captive breeding unless populations are large enough or protected in their habitat to ensure that the population will survive. In order to draw attention to rare and diminishing species the *Red data book* published by the IUCN lists mammals (and other species) by categories of rareness and their status is continually updated (see Smith, 1976). Thus 'endangered species' are in immediate danger of extinction and will not survive if the causal factor now at work continues to operate; examples are the North American black-footed ferret, *Mustela nigripes* which relies on prairie dogs, *Cynomys ludovicianus* as prey and which has declined following poison campaigns against the prairie dog, and the Arabian oryx, *Oryx leucoryx* which has declined in Middle Eastern desert areas following hunting and warfare. 'Vulnerable' or 'threatened' species are still abundant in parts of their range but threatened because of their decline in numbers; they will move into the endangered category in the near future if the causal factors now operating continue. 'Rare' species are nowhere endangered or threatened but are at risk because of low numbers.

Many species have become extinct in the past and before captive breeding is undertaken with the intention of possible reintroduction the reasons for the decline must be properly understood. The captive breeding programmes of zoological gardens and wildlife parks have become prominent in their publicity and reasons for existence but, so far, successes in this field leading to reintroduction have not been common. It is important that habitat conservation and the eventual reintroduction of the species is also an objective alongside captive breeding (Brambell, 1977), unless the restocking of zoos from captive populations is the only possible objective.

Captive breeding has succeeded in a number of mammalian species (Table 8.1) but many others are thought to be capable of sustaining themselves only with further recruitment from the wild (Pinder and Barkham, 1978). Père David's deer (*Elaphurus davidianus*) was brought back to Britain from China in the 1860s and has increased in park and zoo herds while it became extinct in the wild; it is now a suitable species for reintroduction. Przewalski's horse (*Equus przewalski*) from Mongolia has a similar history to Père David's deer but reintroduction is uncertain; however, the Arabian oryx has been successfully bred in Arizona from a handful of captive and captured individuals and is now in the process of reintroduction in Oman. Perhaps the most famous captive breeding programme and reintroduction of a mammal is that of the European bison (*Bison bonasus*). This species was exterminated in the wild in the 1920s, leaving a number in private collections and 40 of these were entered into a pedigree book held by the International Society for the Protection of the European bison founded in 1923 (Fitter, 1967). By 1939 numbers had reached 100 and by 1952 two bulls, two cows and a calf were released in the Bialowieza forest in Poland; numbers have now reached 'carrying capacity' in the forest and culling has been necessary to prevent overpopulation.

Many sponsors of captive breeding must be congratulated for saving a species from extinction but problems still exist in maintaining the stock in captivity. This is because the nature of the species may be changed by captive breeding and these changes may not necessarily be obvious before returning representatives to the wild. The species which are endangered are likely to be predators, intolerant of habitat change, valuable, hunted, intolerant of man and have a long gestation and/or weaning period. Once these problems have been overcome and the species reproduces in captivity, the side-effects of selective breeding with small populations may become obvious; these are likely to be infertility and low genetic variation (Kear, 1977). In Prezwalski's horse captive mares produce fewer foals than less inbred ones and there is a general decline in fertility and longevity as well as the loss of some typical characteristics. In a case study of captive ungulates Ralls *et al.* (1979) showed that 15 species showed higher juvenile mortality in inbred than in non-inbred young. There are, however, large,

Table 8.1 List of the twenty-six rare mammal species identified as having self-sustaining populations in zoos (from Pinder and Barkham, 1978).

Vernacular name	Linnean name
MARSUPIALIA	
Parma wallaby	*Macropus parma*
PRIMATES	
Mongoose lemur	*Lemur mongoz mongoz*
Lion-tailed macaque	*Macaca silenus*
CARNIVORA	
Wolf	*Canis lupus*
African wild dog	*Lycaon pictus*
Bengal tiger	*Panthera tigris tigris*
Siberian tiger	*Panthera tigris altaica*
Sumatran tiger	*Panthera tigris sumatrae*
Leopard	*Panthera pardus*
North China leopard	*Panthera pardus japonensis*
Jaguar	*Panthera onca*
PERISSODACTYLA	
Przewalski's horse	*Equus przewalskii*
Onager	*Equus hemionus onager*
Kulan	*Equus hemionus kulan*
ARTIODACTYLA	
Pygmy hippopotamus	*Choeropsis liberiensis*
Swamp deer	*Cervus duvauceli*
Formosan sika	*Cervus nippon taiouanus*
Père David's deer	*Elaphurus davidianus*
Banteng	*Bos banteng*
European bison	*Bison bonasus*
Lechwe waterbuck	*Kobus leche*
Arabian oryx	*Oryx leucoryx*
Scimitar-horned oryx	*Oryx dammah*
Addax	*Addax nasomaculatus*
Black wildebeest	*Connochaetes gnou*
Arabian gazelle	*Gazella gazella arabica*

apparently healthy populations, such as the laboratory and pet golden hamster (*Mesocricetus auratus*) which started from one pair.

In contemplating reintroduction many factors should be taken into account (Perry, 1979). These include whether the correct foods are available in the new habitat, whether toxic plants will be recognized, whether the original causes of depletion have persisted, whether the introduced animals have the necessary antibodies to combat likely diseases, whether parents are able to protect their young from predators, and whether the introduced population is large enough to allow successful social organization and reproduction. As an example of the last problem, the North American pronghorn (*Antilocapra americana*) introduction into the south-west of the USA needed 30–50 individuals to succeed.

Many aspects of population and reproductive biology need to be understood for successful breeding and reintroduction but population models may also play a part in the understanding of the likely success of reintroduction. Estimates of the minimum number in the population required to withstand general uncertainties may be simulated if enough data are available (Shaffer, 1981). These uncertainties may be 'demographic stochasticity', coming from chance events in the survival and/or reproductive success of a finite number of individuals; 'environmental stochasticity' arising from temporal variation in habitat parameters and in the populations of other species which are competitors, predators, parasites or diseases; 'natural catastrophes' which occur randomly, such as fire, flood or drought, and lastly, 'genetic stochasticity', the changes in gene frequencies due to the founder effect, when the frequencies in the founding individuals are different from the normal gene frequencies found in the ancestral population, due to random fixation of gene frequencies and inbreeding. It is suggested that ways of estimating the population size necessary for long-term survival following a reintroduction might be (a) by experiment, which is usually impossible, (b) by assessing the size and rates of colonization and extinction of island populations with different areas of suitable habitat, which is not often possible, and may not take into account the habitat variability present or the changes in population characteristics associated with different habitats, and (c) theoretical models of population longevity which are criticized because of their often unnatural assumptions. Thus simulation might be favoured as a good alternative to the other methods. Again habitat and population factors need to be taken into account but given the full range of population parameters such as age and sex-specific mortality and fecundity rates, age structure, sex ratios, dispersal and the relationships of these variables to density, simulations can be carried out to assess, for a given probability level, the population size which is likely to lead to survival for a specified number of years. Variables such as those listed above have been used in a simulation of the grizzly bear (*Ursus arctos*) population by Shaffer (1978, quoted in Shaffer, 1981) in the Yellowstone National Park where natural catastrophes were considered unimportant and genetic stochasticity was unknown. Thus the simulation allowed the use of environmental and demographic variables only and suggested that less than 30–70 bears (depending on population characteristics) occupying 2500–7400 km^2 (depending on habitat quality) have less than a 95% chance of surviving 100 years. This is a useful, if sobering, result and similar data could be usefully derived for other species.

APPENDIX

The classification of mammals

The following is a classification of the mammals after Corbet and Hill (1980). The orders (bold) are followed by the number of species within them and divided into families and their geographical distribution. Representative species and genera (where mentioned without classification in the foregoing chapters) are listed under each family name.

Monotremata: monotremes, 3
  Tachyglossidae, spiny echidnas; Australia, New Guinea
  Ornithorhynchidae, platypus; Australia, Tasmania
Marsupialia: marsupials, *c*.254
  Didelphidae, opossums; South, Central and North America
    Didelphis virginiana American opossum
  Caenolestidae, shrew-opossums; Andes
  Microbiotheriidae, colocolos; Chile
  Dasyuridae, marsupial mice; Australia, New Guinea
    Antechinus stuarti, brown antechinus
  Myrmecobiidae, numbat; Australia
  Thylacinidae, Tasmanian wolf (extinct?)
  Notoryctidae, marsupial mole; Australia
  Peramelidae, bandicoots; Australia, New Guinea
  Thylacomyidae, rabbit-bandicoots; Australia
  Phalangeridae, phalangers; Australia, New Guinea
  Burramyidae, pygmy possums; Australia, New Guinea
  Petauridae, gliding phalangers; Australia, New Guinea
  Macropodidae, kangaroos, wallabies; Australia, New Guinea
    Setonyx brachyurus, quokka
    Macropus (Protemnodon) eugenii, tammar wallaby
    M. (Megaleia) rufus, red kangaroo
  Phascolarctidae, koala; Australia
  Vombatidae, wombats; Australia
  Tarsipedidae, honey possum; Australia
Edentata: edentates, 29
  Myrmécophagidae, American anteater; South and Central America
  Bradypodidae, sloths; Central America
  Dasypodidae, armadillos; America
Insectivora: insectivores, *c*. 344
  Solenodontidae, solenodons; Cuba, Hispaniola
  Tenrecidae, tenrecs, otter shrews; Madagascar, South and Central Africa
  Chrysochloridae, golden moles; Africa
  Erinaceidae, hedgehogs, moonrats; Europe, Africa

Erinaceus europaeus, hedgehog
Soricidae, shrews; Eurasia, Africa, America
Sorex araneus, Eurasian common shrew
S. vagrans, vagrant shrew
Blarina brevicauda, northern short-tailed shrew
Neomys fodiens, water shrew
Talpidae, moles, desmans; Eurasia, North America
Scalopus spp., American moles
Talpa europaea, European mole
Scandentia: *c.* 16
Tupaiidae, tree shrews; South East Asia
Dermoptera: 2
Cyncephalidae, flying lemurs; South East Asia
Chiroptera: bats *c.* 950
Pteropodidae, fruit bats; Africa, South Asia, Australia
Pteropus spp., flying foxes
Hypsignathus monstrosus, hammer-headed fruit bat
Eidolon helvum, straw-coloured fruit bat
Miniopterus spp., long-fingered bats
Rhinopomatidae, mouse-tailed bats; North Africa, South Asia
Emballonuridae, sheath-tailed bats; World tropics
Craseonycteridae, hog-nosed bat; South Thailand
Nycteridae, slit-faced bats; Africa, South Asia
Megadermatidae, false vampire bats; Africa, South East Asia, Australia
Rhinolophidae, horseshoe bats; Old World, Australia
Rhinolophus spp., horseshoe bats
Hipposideridae, leaf-nosed bats; Africa, Asia, Australia
Noctilionidae, bulldog bats; New World tropics
Mormoopidae, naked-backed bats, South and Central America
Phyllostomatidae, leaf-nosed bats; New World
Artibeus jamaicensis, Jamaican fruit-eating bat
Desmodontidae, vampire bats; America
Natalidae, funnel-eared bats; South and Central America
Natalus stramineus, Mexican funnel-eared bat
Furipteridae, smoky bats; South and Central America
Thyropteridae, disc-winged bats; South and Central America
Myzopodidae, sucker-footed bat; Madagascar
Vespertilionidae, vespertilionid bats; World
Pipistrellus pipistrellus, pipistrelle bat
Mystacinidae, short-tailed bat; New Zealand
Molossidae, free-tailed bats; World
Primates: *c.* 179
Cheirogaleidae, dwarf lemurs; Madagascar
Lemuridae, large lemurs; Madagascar
Indriidae, leaping lemurs; Madagascar
Daubentonidae, aye-aye; Madagascar
Lorisidae, lorises, bushbabies, South East Asia, Africa
Tarsiidae, tarsiers; South East Asia
Callithricidae, marmosets, tamarins; South and Central America
Cebidae, New World monkeys; South and Central America
Callicebus spp., titis
Cercopithidae, Old World monkeys; Africa, South East Asia
Miopithecus talapoin, talapoin

Pan troglodytes, chimpanzee
Papio spp., baboons
Cercocebus spp., mangabeys
Presbytis entellus, langur
Pongidae, apes; West and Central Africa, South East Asia
Pongo pygmaeus, orang-utan
Hominidae, man; World
Carnivora: carnivores *c.* 240
Canidae, dogs, foxes; Eurasia, America, Africa
Vulpes vulpes, red fox
Alopex lagopus, Arctic fox
Speothos venaticus, bush dog
Canis lupus, wolf
C. latrans, coyote
Ursidae, bears; Eurasia, North America
Ursus arctos, grizzly bear
Procyonidae, racoons; America
Ailuropodidae, pandas; Asia
Mustelidae, weasels etc.; America, Eurasia, Africa
Mustela nivalis, weasel
M. erminea, stoat (North American weasel)
M. putorius, polecat
M. vison, American mink
M. nigripes, black-footed ferret
Martes spp., martens
Lutra spp., otters
Meles meles, Eurasian badger
Spilogale putorius, spotted skunk
Viverridae, civets, mongooses; South Eurasia, Africa, Madagascar
Hyaenidae, hyaenas; Africa, South West Asia
Felidae, cats; America, Eurasia, Africa
Puma concolor, mountain lion
Panthera leo, lion
Felis lynx, lynx
Felis catus, domestic cat
Pinnipedia: seals, sealions, *c.* 34
Otariidae, eared seals, sealions; Southern Hemisphere, North Pacific
Callorinus ursinus, northern fur seal
Odobenidae, walrus; Arctic
Phocidae, earless seals; Southern Hemisphere, North Pacific
Mirounga angustirostris, northern elephant seal
Pagophilus groenlandicus, Atlantic harp seal
Halichoerus grypus, grey seal
Cetacea: whales, dolphins, *c.* 76
Platanistidae, river dolphins; South America, South East Asia
Delphinidae, dolphins; World
Phocoenidae, porpoises; North and South Temperate, Arctic
Monodontidae, white whale, narwhale; Arctic
Physeteridae, sperm whales; World
Ziphiidae, beaked whales; World
Eschrichtidae, grey whale; North Pacific
Balaenopteridae, rorquals; World
Balaenoptera borealis, sei whale

B. physalis, fin whale
B. musculus, blue whale
Balaenidae, right whales; World
Sirenia: sea cows, 4
Dugongidae, dugong; Indian, tropical Atlantic coasts
Trichechidae, manatees; tropical Atlantic coasts
Proboscidea: 2
Elephantidae, elephants; Africa, South East Asia
Loxodonta africana, African elephant
Perissodactyla: odd-toed ungulates, 17
Equidae, horses etc.; Africa, West and Central Asia
Equus przewalski, Przewalski's horse
Equus burchelli, common zebra
Tapiridae, tapirs; South and Central America, South East Asia
Rhinocerotidae, rhinoceroses; Africa, South East Asia
Hyracoidea: *c.* 5
Procaviidae, hyraxes; Africa, Arabia
Procavia capensis, rock hyrax
Tubulidentata: 1
Orycteropodidae, aardvark; Africa
Artiodactyla: even-toed ungulates, *c.* 184
Suidae, pigs; Eurasia, Africa
Phacochoerus aethiopicus, warthog
Tayassuidae, peccaries; America
Hippopotamidae, hippos; Africa
Camelidae, camels, llamas; South America, Asia, North America
Camelus spp., camels
Lama guanicoe, llama
Lama glama, domestic llama
Lama pacos, alpaca
Tragulidae, mouse-deer; West and Central Africa, South East Asia
Moschidae, musk deer; East Asia
Cervidae, deer; America, Eurasia
Dama dama, fallow deer
Capreolus capreolus, roe deer
Cervus canadensis, wapiti
Cervus elaphus, red deer
Odocoileus virginianus, white-tailed deer
O. hemionus, black-tailed deer
Elaphurus davidianus, Père David's deer
Rangifer tarandus, reindeer, caribou
Alces alces, moose, elk
Giraffidae, giraffe, okapi; Africa
Giraffa camelopardalis, giraffe
Antilocapridae, pronghorn; North and Central America
Antilocapra americana, American pronghorn
Bovidae, cattle, antelope, sheep, goats; North America, Africa, Eurasia
Gazella spp., gazelles
Madoqua kirki, Kirk's dik-dik
Kobus (Adenota) kob, kob
K. ellipsiprymnus (defassa), waterbuck
Damaliscus korrigum, topi
Connochaetes taurinus, wildebeeste

Alcelaphus buselaphus cokei, Coke's hartebeeste
Ovis aries, Soay sheep, domestic sheep
Capra hircus, goat
Syncerus caffer, African buffalo
Bison bonasus, European bison
Oryx leucoryx, Arabian oryx
Ovibos moschatus, musk ox
Rupicapra rupicapra, chamois

Pholidota: 7
Manidae, pangolins; Africa, South Asia

Rodentia: rodents, *c*. 1591
Aplodontidae, mountain beaver; North America
Sciuridae, squirrels, marmots; Eurasia, Africa, America
Sciurus carolinensis, grey squirrel
S. vulgaris, Eurasian red squirrel
Tamaisciurus hudsonicus, American red squirrel
Marmota flaviventris, yellow-bellied marmot
Spermophilus (Citellus), ground squirrels
Cynomys spp., prairie dogs
Tamias (Eutamias) spp., chipmunks
Geomyidae, pocket gophers; America
Geomys spp., pocket gophers
Thomomys talpoides, pocket gophers
Heteromyidae, pocket mice; America
Perognathus spp., pocket mice
Castoridae, beavers; North America, Eurasia
Castor fiber, Eurasian beaver
Anomaluridae, scaly-tailed squirrels; West and Central Africa
Pedetidae, spring hare; South Africa
Muridae, mice, rats, voles, gerbils, hamsters; World
Mus musculus, house mouse
Oryzomys palustris, rice rat
Sigmodon hispidus, cotton rat
Apodemus sylvaticus, wood mouse
A. flavicollis, yellow-necked mouse
Rattus norvegicus, brown rat
Microtus agrestis, field vole (short-tailed vole)
M. arvalis, common vole
M. montanus, montane vole
M. ochrogaster, prairie vole
M. pennsylvanicus, meadow vole
M. breweri, meadow vole
M. townsendii, Townsend vole
M. californicus, California vole
Peromyscus maniculatus, deer mouse
P. leucopus, white-footed mouse
P. polionotus, old-field mouse
P. gossipinus, cotton mouse
Dicrostonyx spp., collared lemmings
Lemmus spp., brown lemmings
Clethrionomys glareolus, bank vole
Ondatra zibithecus, muskrat
Mesocricetus auratus, golden hamster

Phodopus sungorus, Djungarian hamster
Onychomys torridus, grasshopper mouse
Ochrotomys nuttalli, golden mouse
Neotoma floridana, Eastern woodrat
Gliridae, dormice; Eurasia, Africa
Seleviniidae, desert dormouse; Asia
Zapodidae, jumping mice; North Eurasia, North America
Dipodidae, jerboas; Africa, Central Asia
Hystricidae, Old World porcupines; Africa, South Asia
Erithizontidae, New World porcupines; America
Caviidae, Guinea pigs; South America
Cavia porcellus, Guinea pig
Galea musteloides, cuis
Hydrochoeridae, capybara; South America
Dinomyidae, pacarana; South America
Dasyproctidae, agoutis, pacas; South and Central America
Chinchillidae, chinchillas, viscachas; South America
Lagostomus maximus, plains viscacha
Chinchilla laniger, chinchilla
Capromyidae, coypus, hutias, South and Central America
Myocastor coypus, coypu
Octodontidae, degus, coruru; South America
Ctenomyidae, tuco-tucos; South America
Abrocomidae, chinchilla rats; South America
Echimyidae, America spiny rats; South and Central America
Thyronomyidae, cane rats; Africa
Petromyidae, African rock rat; Africa
Bathyergidae, African mole-rats; Africa
Ctenodactylidae, gundis; Africa
Lagomorpha: lagomorphs *c*. 54
Ochotonidae, pikas; North and Central Asia, North America
Leporidae, rabbits, hares; Eurasia, Africa, North and Central America
Oryctolagus cuniculus, rabbit
Lepus europaeus, hare
L. americanus, snowshoe hare
L. timidus, mountain hare
L. californicus, black-tailed jackrabbit
Macroscelidea: *c*. 15
Macroscelididae, elephant-shrews; Africa
Elephantulus myurus, Transvaal elephant-shrew
E. rufescens, rufous elephant shrew
Rhynchocyon chrysopygus, yellow-rumped elephant shrew

References

Recommended further reading

Eisenberg, J.F. (1981). *Mammalian Radiation: An Analysis of Trends in Evolution, Adaptation and Behaviour*, Athlone Press.
Young, J.Z. (1975). *Life of Mammals: Their Anatomy and Physiology*, Oxford University Press.

Abramsky, Z. and Tracy, C.R. (1979). Population biology of 'noncycling' population of prairie voles and an hypothesis on the rate of migration in regulating microtine cycles. *Ecology*, **60**, 349–61.
Alexander, R.D. (1974). The evolution of social behaviour. *Annual Review of Ecology and Systematics*, **5**, 325–83.
Alibhai, S.K. and Gipps, J.H.W. (1985). The population dynamics of bank voles. *Symposia of the Zoological Society of London*, **55**, 277–313.
Allen, K.R. (1980). *Conservation and Management of Whales*. Butterworths, London.
Almeida, O.F.X. and Lincoln, G.A. (1984). Reproductive photorefractoriness in rams and accompanying changes in the patterns of melatonin and prolactin secretion. *Biology of Reproduction*, **30**, 143–58.
Altmann, S.A., Wagner, S.S. and Lenington, S. (1977). Two models for the evolution of polygyny. *Behavioral Ecology and Sociobiology*, **2**, 397–410.
Amoroso, E.C. and Marshall, F.H.A. (1960). External factors in sexual periodicity. In *Marshall's Physiology of Reproduction Vol. III, Part II.*, 3rd edition, pp. 707–831. A.S. Parkes (ed.). Longmans, Green & Co., London.
Amoroso, E.C. and Matthews, J.H. (1951). The growth of the grey seal (*Halichoerus grypus*) from birth to weaning. *Journal of Anatomy*, **85**, 427–8.
Anderson, D.C., Armitage, K.B. and Hoffman, R.S. (1976). Socioecology of marmots: female reproductive strategies. *Ecology*, **57**, 552–60.
Anderson, G.B. (1977). Fertilization, early development and embryo transfer. In *Reproduction in Domestic Animals*, 3rd edition, pp. 285–314. H.H. Cole and P.T. Cupps (eds). Academic Press, London.
Anderson, J.L. (1975). *Phenotypic correlates among relatives and variability in reproductive performance in populations of the vole,* Microtus townsendii. PhD Thesis, Universty of British Columbia. (Quoted by Krebs, 1978).
Anderson, P.K. (1970). Ecological structure and gene flow in small mammals. *Symposia of the Zoological Society of London*, **26**, 299–325.
Anderson, R.M. and May, R.M. (1979). Population biology of infectious diseases: Part I. *Nature, London*, **280**, 361–7.
Andrewartha, H.G. and Birch, L.C. (1954). *The Distribution and Abundance of Animals*. University of Chicago Press, Chicago.

Andrews, R.V., Belknap, R.W., Southard, J., Lorincz, M. and Hess, S. (1972). Physiological, demographic, and pathological changes in wild Norway rat populations over an annual cycle. *Comparative Biochemistry and Physiology*, **41A**, 149–65.

Armitage, K.B. (1965). Vernal behaviour of the yellow-bellied marmot (*Marmota flaviventris*). *Animal Behaviour*, **13**, 59–68.

Armitage, K.B. (1974). Male behaviour and territoriality in the yellow-bellied marmot. *Journal of Zoology, London*, **172**, 233–65.

Aron, C. (1979). Mechanisms of control of the reproductive function by olfactory stimuli in female mammals. *Physiological Reviews*, **59**, 229–84.

Asdell, S.A. (1964). *Patterns of Mammalian Reproduction*, 2nd edition. Cornell University Press, Ithaca, New York.

Ashdown, R.R. and Hancock, J.L. (1974). Functional anatomy of male reproduction. In *Reproduction in Farm Animals*, pp. 3–23. E.S.E. Hafez (ed.). Lea & Febiger, Philadelphia.

Axelrod, J. (1974). The pineal gland: A neurochemical transducer. *Science, New York*, **184**, 1341–8.

Baker, J.R. (1938a). Evolution of breeding seasons. In *Evolution: Essays on Aspects of Evolutionary Biology Presented to Professor E.S. Goodrich*, pp. 161–77. G.R. de Beer (ed.). Oxford University Press, (Clarendon), London.

Baker, J.R. (1938b). The relation between latitude and breeding seasons in birds. *Proceedings of the Zoological Society of London (A)*, **108**, 557–82.

Baker, J.R. and Ranson, R.M. (1932). Factors affecting the breeding of the field mouse (*Microtus agrestis*). Part I – light. *Proceedings of the Royal Society of London (B)*, **110**, 313–22.

Baker, T.G. (1982). Oogenesis and ovulation. In *Reproduction in Mammals: 1. Germ Cells and Fertilization*, pp. 17–45. C.R. Austin and R.V. Short (eds). Cambridge University Press, Cambridge.

Baird, D.T. (1978). Local utero-ovarian relationships. In *Control of Ovulation*, pp. 217–33. D.B. Crighton, N.B. Haynes, G.R. Foxcroft, and G.E. Lamming (eds). Butterworths, London.

Baird, D.T. (1983). The ovary. In *Reproduction in Mammals: 3. Hormonal Control of Reproduction*, 2nd edition., pp. 91–114. C.R. Austin and R.V. Short (eds). Cambridge University Press, Cambridge.

Baird, D.T., Baker, T.G., McNatty, K.P. and Neal, P. (1975). Relationship between the secretion of the corpus luteum and the length of the follicular phase of the overian cycle. *Journal of Reproduction and Fertility*, **45**, 611–19.

Barkalow, F.S., Hamilton, R.B. and Soots, R.F. (1970). The vital statistics of an unexploited gray squirrel population. *Journal of Wildlife Management*, **34**, 489–500.

Barnett, S.A. (1962). Total breeding capacity of mice at two temperatures. *Journal of Reproduction and Fertility*, **4**, 327–5.

Barnett, S.A. and Coleman, E.H. (1959). The effect of low environmental temperature on the reproductive cycle of female mice. *Journal of Endocrinology*, **10**, 232–40.

Bartholomew, G.A. (1952). Reproductive and social behavior in the northern elephant seal. *University of California Publications in Zoology*, **47**, 369–472.

Bartholomew, G.A. (1970). A model of the evolution of pinneped polygyny. *Evolution*, **24**, 546–59.

Bartholomew, G.A. and Hoel, P.G. (1953). Reproductive behavior of the Alaska fur seal, *Callorhinus ursinus*. *Journal of Mammalogy*, **34**, 417–36.

Batzli, G.O. (1950). The role of small mammals in arctic ecosystems. In *Small Mammals: Their Productivity and Population Dynamics, IBP 5*, pp. 243–68. F.B.

Golley, K. Petrusewicz and L. Ryszkowski (eds). Cambridge University Press, Cambridge.

Batzli, G.O. and Pitelka, F.A. (1975). Vole cycles – test of another hypothesis. *American Naturalist*, **109**, 482–7.

Beach, F.A. (1976). Sexual attractivity, proceptivity, and receptivity in female mammals. *Hormones and Behaviour*, **7**, 105–38.

Beacham, T.D. (1979). Selectivity of avian predation in declining populations of the vole, *Microtus townsendii*. *Canadian Journal of Zoology*, **57**, 1767–72.

Bedford, Duke of, and Marshall, F.H.A. (1942). On the incidence of the breeding season in mammals after transference to a new latitude. Proceedings of the Royal Society of London (B), **130**, 396–9.

Bedford, J.M. (1983). Fertilization. In *Reproduction in Mammals: 1. Germ Cells and Fertilization*, 2nd edition. pp. 128–63. C.R. Austin and R.V. Short (eds). Cambridge University Press, Cambridge.

Bekoff, M. (1981). Vole population cycles: kin selection or familiarity? *Oecologia*, **48**, 131.

Bell, G. (1971). On breeding more than once. *American Naturalist*, **110**, 57–77.

Berger, P.J., Negus, N.C., Sanders, E.H. and Gardner, P.D. (1981). Chemical triggering of reproduction in *Microtus montanus*. *Science, New York*, **214**, 69–70.

Berger, P.J., Sanders, E.H., Gardner, P.D. and Negus, N.C. (1977). Phenolic plant compounds functioning as reproductive inhibitors in *Microtus montanus*. *Science, New York*, **195**, 575–7.

Bergeron, J.-M. (1980). Importance des plantes toxiques dans le régime alimentaire de *Microtus pennsylvanicus* à deux étapes opposées de leur cycle. *Canadian Journal of Zoology*, **58**, 2230–8.

Bergerud, A.T., Nolan, M.J., Curnew, K. and Mercer, W.E. (1983). Growth of the Avalon peninsula, Newfoundland caribou herd. *Journal of Wildlife Management*, **47**, 989–98.

Bergerud, A.T. Wyett, W. and Snider, B. (1983). The role of wolf predation in limiting a moose population. *Journal of Wildlife Management*, **47**, 977–88.

Bernstein, I.S. (1976). Dominance, aggression and reproduction in primate societies. *Journal of Theoretical Biology*, **60**, 459–72.

Berry, R.J. (1981). Population dynamics of the house mouse. *Symposia of the Zoological Society of London*, **47**, 395–425.

Bertram, B.C.R. (1973). Lion population regulation. *East African Wildlife Journal*, **11**, 215–25.

Bertram, B.C.R. (1975). Social factors influencing reproduction in wild lions. *Journal of Zoology, London*, **117**, 463–82.

Bertram, B.C.R. (1976). Kin selection in lions and in evolution. In *Growing Points in Ethology*, pp. 281–301. P.P.G. Bateson and R.A. Hinde (eds). Cambridge University Press, Cambridge.

Beuchner, H.K. (1974). Implications of social behaviour in the management of Uganda kob. In *The Behaviour of Ungulates and its Relation to Management, IUCN Series No. 24*, pp. 853–70. V. Geist and F. Walther (eds). IUCN, Morges.

Birdsall, D.A. and Nash, D. (1973). Occurrence of multiple insemination of females in natural populations of deer mice (*Peromyscus maniculatus*). *Evolution*, **27**, 106–110.

Birney, E.C., Grant, W.E. and Baird, D.D. (1976). Importance of vegetative cover to cycles of *Microtus* populations. Ecology, **57**, 1043–51.

Blandau, R.J. (1973). Gamete transport in the female mammal. In *Handbook of Physiology, Section 7: Endocrinology, Vol. II, Part II*, pp. 153–63. R.O. Greep and E.B. Astwood (eds). American Physiological Society, Washington. Williams and Wilkins, Baltimore, Maryland.

Boer, P.J. Den (1968). Spreading of risk and stabilization of animal numbers. *Acta Biotheoretica*, **18**, 165–94.

Bonner, W.N. (1980). *Whales*. Blandford Press, Poole, Dorset.

Boonstra, R. and Krebs, C.J. (1977). A fencing experiment on a high-density population of *Microtus townsendii*. *Canadian Journal of Zoology*, **55**, 1166–75.

Bowen, W.D., Capstick, C.K. and Sergeant, D.E. (1981). Temporal changes in the reproductive potential of female harp seals (*Pagophilus groenlandicus*). *Canadian Journal of Fisheries and Aquatic Science*, **38**, 495–503.

Bradbury, J.W. (1977). Lek mating behaviour in the hammer-headed bat. *Zeitschrift für Tierpsychology*, **45**, 225–55.

Bradbury, R.B. and White, D.E. (1954). Estrogens and related substances in plants. *Vitamins and Hormones*, **12**, 207–33.

Bradshaw, G.V.R. (1962). Reproductive cycle of the California leaf-nosed bat, *Macrotus californicus*. *Science, New York*, **136**, 645–6.

Braithwaite, R.W. and Lee, A.K. (1979). A mammalian example of semelparity. *American Naturalist*, **113**, 151–55.

Brambell, F.W.R. (1956). Ovarian changes. In *Marshall's Physiology of Reproduction, Vol. I, part I*, 3rd edition, pp. 397–542. A.S. Parkes (ed.). Longmans, Green & Co., New York.

Brambell, F.W.R. and Rowlands, I.W. (1936). Reproduction of the bank vole (*Evotomys glareolus* Schreber). I. The oestrous cycle of the female. *Philosophical Transactions of the Royal Society of London (B)*, **226**, 71–97.

Brambell, M.R. (1977). Reintroduction. In *International Zoo Yearbook, 17*, pp. 112–6. P.J.S. Olney, (ed.). Zoological Society of London, London.

Bramley, P.S. (1970). Territoriality and reproductive behaviour of roe deer. *Journal of Reproduction and Fertility, Supplement*, **11**, 43–70.

Breed, W.G. and Clarke, J.R. (1970). Effect of photoperiod on ovarian function in the vole, *Microtus agrestis*. *Journal of Reproduction and Fertility*, **23**, 189–92.

Broekhuizen, S. (1975). The position of the wild rabbit in the life system of the European hare. *Proceedings of the International Congress of Game Biologists, Lisbon*, **12**, 75–79.

Bronson, F.H. (1974). Pheromonal influences on reproductive activities in rodents. In *Pheromones*, pp. 344–65. M. Birch (ed.). North Holland Publishing Co., Amsterdam.

Bronson, F.H. (1979). The reproductive ecology of the house mouse. *Quarterly Review of Biology*, **54**, 265–99.

Bronson, F.H. and Caroom, D. (1971). Preputial gland of the male mouse: attractant function. *Journal of Reproduction and Fertility*, **25**, 279–82.

Bronson, F.H. and Maruniak, J.A. (1975). Male-induced puberty in female mice: evidence for a synergistic action of social cues. *Journal of Reproduction and Fertility*, **13**, 94–8.

Bronson, F.H., Eleftheriou, B.E. and Dezell, H.E. (1969). Strange male pregnancy block in deermice: prolactin and adrenocortical hormones. *Biology of Reproduction*, **1**, 302–6.

Brown, J.L. (1975). *The Evolution of Behavior*. Norton, New York.

Brown-Grant, K., Davidson, J.M. and Greig, F. (1973). Induced ovulation in albino rats exposed to constant light. *Journal of Endocrinology*, **57**, 7.

Brown, R.E. and Macdonald, D.W. (eds) (1985). *Social Odours in Mammals Vols 1 and 2*. Clarendon Press, Oxford.

Bruce, H.M. (1959). An exteroceptive block to pregnancy in the mouse. *Nature. London*, **184**, 105.

Bryant, J.P. and Kuropat, P.J. (1980). Selection of winter forage by subarctic browsing vertebrates: the role of plant chemistry. *Annual Review of Ecology and*

Systematics, **11**, 261–85.
Bunnell, F.L. (1982). The lambing period of mountain sheep: synthesis, hypothesis and tests. *Canadian Journal of Zoology*, **60**, 1–14.
Bunnell, F.L., Eastman, D.S. and Peek, J.M. (eds) (1983). *Natural Regulation of Wildlife Populations (March 10, 1978)*. Forest, Wildlife and Range Experimental Station, University of Idaho, Moscow. (Not seen in the original).
Bünning, E. (1936). Die endogenie Tagesrhythmik als Grundlage der Photoperiodischen Reaktion. *Bereicht der Deutschen botanischen Gasellschaft*, **54**, 590–607.
Bünning, E. (1967). *The Physiological Clock*, 2nd edition. English Universities Press, London.
Burt, W.H. (1943). Territoriality and home range concepts as applied to mammals. *Journal of Mammalogy*, **24**, 346–52.
Calhoun, J. (1949). A method for self-control of population growth among mammals living in the wild. *Science, New York*, **109**, 333–5.
Cameron, R.D.A. and Blackshaw, A.W. (1980). The effect of elevated ambient temperature on spermatogenesis in the boar. *Journal of Reproduction and Fertility*, **59**, 173–9.
Campbell, B. (ed.) (1972). *Sexual Selection and the Descent of Man*. Heinemann, London.
Carl, E.A. (1971). Population control in arctic ground squirrels. *Ecology*, **52**, 396–413.
Carr, W.J., Loeb, L.S. and Dissinger, M.L. (1965). Responses of rats to sex odors. *Journal of Comparative Physiology and Psychology*, **59**, 370–7.
Carrick, F.N. and Setchell, B.P. (1977). The evolution of the scrotum. In *Reproduction and Evolution*, pp. 165–70. J.H. Calaby and C.H. Tyndale-Biscoe (eds). Australian Academy of Science, Canberra.
Carter, C.S., Getz, L.L., Gavish, L., McDermott, J.L. and Arnold, P. (1980). Male-related pheromones and the activation of female reproduction in the prairie vole (*Microtus ochrogaster*). *Biology of Reproduction*, **23**, 1038–45.
Case, T.J. (1978). On the evolution and adaptive significance of postnatal growth rates in the terrestrial vertebrates. *Quarterly Review of Biology*, **53**, 243–82.
Caughley, G. (1976a). Wildlife management and the dynamics of ungulate populations. *Applied Biology*, **1**, 183–246.
Caughley, G. (1976b). Plant-herbivore systems. In *Theoretical Ecology*, pp. 94–113. R.M. May (ed.). Blackwell Scientific Publications, Oxford.
Caughley, G. (1976c). The elephant problem – an alternative hypothesis. *East African Wildlife Journal*, **14**, 265–83.
Caughley, G. (1977). *Analysis of Vertebrate Populations*. John Wiley & Sons Ltd., Chichester.
Chalmers, N. (1979). *Social Behaviour in Primates*. Edward Arnold, London.
Champlin, A.K. (1971). Suppression of oestrus in grouped mice: the effects of various densities and the possible nature of the stimulus. *Journal of Reproduction and Fertility*, **27**, 233–41.
Champlin, A.K., Dorr, D.L. and Gates, A.H. (1973). Determining the stage of the oestrous cycle in the mouse by the appearance of the vagina. *Biology of Reproduction*, **8**, 491–4.
Chang, M.C. and Hunter, R.H.F. (1975). Capacitation of mammalian sperm: biological and experimental aspects. In *Handbook of Physiology, Section 7. Endocrinology Vol V*, pp. 339–51. R.O. Greep and E.B. Astwood (eds). American Physiological Society, Washington.
Chanin, P.R.F. and Jefferies, D.J. (1978). The decline of the otter *Lutra lutra* L. in Britain: an analysis of hunting records and discussion of causes. *Biological Journal of the Linnaean Society*, **10**, 305–28.

Chapman, D.I. (1977). Roe Deer *Capreolus capreolus*. In *The Handbook of British Mammals* 2nd edition, pp. 437–46. G.B. Corbert, and H.N. Southern (eds). Blackwell Scientific Publications, Oxford.

Chapman, D.I. and Chapman, N.G. (1975). *Fallow Deer: Their History, Distribution and Biology*. Terence Dalton, Lavenham.

Charnov, E.L. and Finerty, J.P. (1980). Vole population cycles: a case for kin-selection? *Oecologia*, **45**, 1–2.

Charnov, E.L. and Krebs, J.R. (1974) On clutch size and fitness. *Ibis*, **116**, 217–9.

Charnov, E.L. and Schaffer, W.M. (1973). Life history consequences of natural selection. Cole's result revisited. *American Naturalist*, **107**, 791–3.

Chitty, D. (1960). Population processes in the vole and their relevance to general theory. *Canadian Journal of Zoology*, **38**, 99–113.

Chitty, D. (1967). The natural selection of self-regulatory behaviour in animal populations. *Proceedings of the Ecological Society of Australia*, **2**, 51–78.

Christian, J.J. (1950). The adreno-pituitary system and population cycles in mammals. *Journal of Mammalogy*, **31**, 247–59.

Christian, J.J. (1971). Population density and reproductive efficiency. *Biology of Reproduction*, **4**, 248–94.

Christian, J.J. (1980). Endocrine factors in population regulation. In *Biosocial Mechanisms of Population Regulation*, pp. 55–115. M.N. Cohen, R.S. Malpass and H.G. Klein (eds). Yale University press, New Haven.

Christian, J.J. and Davis, D.E. (1964). Endocrines, behavior and population. *Science, New York*, **146**, 1550–60.

Clarke, J.J. and Cummins, J.T. (1982). The temporal relationship between gonadotrophin releasing hormone (GnRH) and luteinizing hormone (LH) secretion in ovariectomized ewes. *Endocrinology*, **111**, 1737–9.

Clarke, J.R. (1972). Seasonal breeding in female mammals. *Mammal Review*, **1**, 217–30.

Clutton-Brock, T.H. and Harvey, P. (1977). Primate ecology and social organisation. *Journal of Zoology*, **183**, 1–39.

Clutton-Brock, T.H., Guinness, F.E. and Albon, S.D. (1982). *Red Deer. Behavior and Ecology of Two Sexes*. Edinburgh University Press, Edinburgh. © By The University of Chicago Press.

Clutton-Brock, T.H., Major, M. and Guinness, F.E. (1985). Population regulation in male and female red deer. *Journal of Animal Ecology*, **54**, 831–46.

Cole, L.C. (1954). The population consequences of life-history phenomena. *Quarterly Review of Biology*, **29**, 103–37.

Cole, F.R. and Batzli, G.O. (1978). Influences of supplemental feeding on a vole population. *Journal of Mammalogy*, **59**, 809–19.

Cohen, J. and McNaughton, D. (1974). Spermatozoa: the probable selection of a small population by the genital tract of the female rabbit. *Journal of Reproduction and Fertility*, **39**, 297–310.

Colby, D.R. and Vandenbergh, J.G. (1974). Regulatory effects of urinary pheromones on puberty in the mouse. *Biology of Reproduction*, **11**, 268–79.

Conaway, C.H. (1971). Ecological adaptation and mammalian reproduction. *Biology of Reproduction*, **4**, 239–47.

Corbet, G.B. and Hill, J.E. (1980). *A World List of Mammalian Species*. British Museum (Natural History), London.

Cowan, D.P. and Garson, P.J. (1985). Variations in the social structure of rabbit populations: causes and demographic consequences. In *Behavioural Ecology: Ecological Consequences of Adaptive Behaviour*, pp. 537–55. R.M. Silby and R.H. Smith (eds). Blackwell Scientific Publications, Oxford.

Cox, C.R. and LeBoeuf, B.J. (1977). Female incitation of male competition: a mechanism in sexual selection. *American Naturalist*, **111**, 317–335.

Crook, J.H., Ellis, J.E. and Goss-Custard, J.D. (1976). Mammalian social systems: structure and function. *Animal Behaviour*, **24**, 261–74.

Crowcroft, W.P. and Rowe, F.P. (1963). Social organization and territorial behaviour in the wild house mouse (*Mus musculus* L.). *Proceedings of the Zoological Society of London*, **140**, 517–31.

Curtis, R.F., Bullentine, J.A., Keverne, E.B., Bonsal, R.W. and Michael, R.P. (1971). Identification of primate sexual pheromones and the properties of synthetic attractants. *Nature, London*, **232**, 396–8.

Daly, M. and Wilson, M. (1978). *Sex, Evolution and Behaviour*. Duxbury Press, Massachusetts.

Darwin, C. (1871). *The Descent of Man and Selection in Relation to Sex*. John Murray, London.

Darling, F. Fraser (1938). *Bird Flocks and the Breeding Cycle*. Cambridge University Press, London.

Deansley, R. (1943). Delayed implantation in the stoat (*Mustela mustela*). *Nature, London*, **151**, 365–6.

Delany, M.J. and Bishop, I.R. (1960). The systematics, life-history and evolution of the bank vole *Clethrionomys* Tilesius in North-West Scotland. *Proceedings of the Zoological Society of London*, **135**, 409–22.

Desjardins, C. (1981). Endocrine signalling and male reproduction. *Biology of Reproduction*, **24**, 1–21.

Dewsbury, D.A. (1972). Patterns of copulatory behaviour in male mammals. *Quarterly Review of Biology*, **47**, 1–33.

Diamond, M. (1970). Intromission pattern and species vaginal code in relation to induction of pseudopregnancy. *Science, New York*, **169**, 995–7.

Dobson, F.S. (1979). An experimental study of dispersal in Californian ground squirrels. *Ecology*, **60**, 1103–9.

Dobzhansky, T.H. (1950). Evolution in the Tropics. *American Scientist*, **38**, 209–221.

Dorrington, J. (1979). Pituitary and placental hormones. In *Reproduction in mammals 7. Mechanisms of Hormone Action*, pp. 53–80. C.R. Austin and R.V. Short (eds). Cambridge University Press, Cambridge.

Doty, R.L. and Dunbar, I.F. (1974). Attraction of beagles to conspecific urine, vaginal and anal sac odors. *Physiology and Behavior*, **12**, 825–33.

Downhower, J.F. and Armitage, K.B. (1971). The yellow-bellied marmot and the evolution of polygamy. *American Naturalist*, **105**, 355–70. The University of Chicago Press.

Drickamer, L.C. (1974). Sexual maturation of female house mice: social inhibition. *Developmental Psychobiology*, **7**, 257–65.

Ducker, M.J., Bowman, J.C. and Temple, A. (1973). The effect of constant photoperiod on the expression of oestrus in the ewe. *Journal of Reproduction and Fertility, Supplement*, **19**, 143–50.

Dukelow, W.R. (1966). Variations in gestation length of mink (*Mustela vison*). *Nature, London*, **211**, 211.

Dunmire, W.W. (1960). An altitudinal survey of reproduction in *Peromyscus maniculatus*. *Ecology*, **41**, 174–82.

Duvall, S.W., Bernstein, I.S. and Gordon, T.P. (1976). Paternity and status in a rhesus monkey group. *Journal of Reproduction and Fertility*, **47**, 25–31.

Dym, M. and Fawcett, D.W. (1970). The blood-testis barrier in the rat and the physiological compartmentation of the seminiferous epithelium. *Biology of Reproduction*, **3**, 308–26.

Eckstein, P. and Zuckerman, S. (1955). Reproduction in mammals. In *Comparative Physiology of Reproduction and Effects of Sex Hormones in Vertebrates*, pp.114–28. I. Chester Jones and P. Eckstein (eds). Memoir of the Endocrinological Society No. 4, Cambridge University Press, London.

Eckstein, P. and Zuckerman, S. (1956). Morphology of the reproductive tract. In *Marshall's Physiology of Reproduction*, 3rd edition pp. 43–155. A.S. Parkes (ed.). Longmans, Green & Co., New York.
Egbunike, G.N. and Elemo, A.O. (1978). Testicular and epididymal sperm reserves of crossbred European boars raised and maintained in the humid tropics. *Journal of Reproduction and Fertility*, **54**, 245–8.
Eleftheriou, B.F., Bronson, F.H. and Zarrow, M.X. (1962). Interaction of olfactory and other environmental stimuli on implantation in the deermouse. *Science, New York*, **137**, 764.
Elliott, J.A. (1976). Circadian rhythms and photoperiodic time measurement in mammals. *Federation Proceedings*, **35**, 2339–46.
Elliott, J.A., Stetson, M.H. and Menaker, M. (1972). Regulation of testis function in golden hamsters: circadian clock measures photoperiodic time. *Science, New York*, **178**, 771–3.
Elliott, P.F. (1975). Longevity and the evolution of polygyny. *American Naturalist*, **109**, 281–7.
Elseth, G.D. and Baumgartner, K.D. (1981). *Population Biology*. Van Nostrand Co., New York.
Eltringham, S.K. (1979). *The Ecology and Conservation of African Mammals*. Macmillan, London.
Elton, C.S. (1942). *Voles Mice and Lemmings: Problems of Population Dynamics*. Oxford University Press, London.
Emlen, S.T. and Oring, L.W. (1977). Ecology, sexual selection and the evolution of mating systems. *Science, New York*, **197**, 215–23.
Emmens, C.W. and Gidley-Baird, A.A. (1977). Evolution of the oestrous cycle. In *Reproduction and Evolution*, pp. 235–43. J.H. Calaby and C.H. Tyndale-Biscoe (eds). Australian Academy of Science, Canberra.
Erickson, B.H. (1966). Development and senescence of the postnatal bovine ovary. *Journal of Animal Science*, **25**, 800–5.
Erlinge, S. (1983). Demography and dynamics of a stoat *Mustela erminea* population in a diverse community of vertebrates. *Journal of Animal Ecology*, **52**, 705–26.
Erlinge, S., Goransson, G., Hansson, L., Högstedt, G., Liberg, O., Nilsson, I.N., Nilsson, T., Schrantz, T. von and Sylven, M.C. (1983). Predation as a regulating factor on small rodent populations in southern Sweden. *Oikos*, , **40**, 36–52.
Errington, P.L. (1946). Predation and vertebrate populations. *Quarterly Review of Biology*, **21**, 144–77, 221–45.
Estes, R.D. (1972). The role of the vomeronasal organ in mammalian reproduction. *Mammalia*, **36**, 315–41.
Everett, J.W. (1961). The mammalian female reproductive cycle and its controlling mechanisms. In *Sex and Internal Secretions*, 3rd edition, pp. 497–555. W.C. Young (ed.). Williams and Wilkins, Baltimore.
Ewer, R.F. (1968). *Ethology of Mammals*. Logo Press Ltd., London.
Ewing, L.L., Davis, J.C. and Zirkin, B.R. (1980). Regulation of testicular function: A spatial and temporal view. In *Reproductive Physiology III*, pp. 41–115. R.O. Greep (ed.). International Review of Physiology. Vol. 22, University Park Press, Baltimore.
Fairbairn, D.J. (1977a). Why breed early? A study of reproductive tactics in *Peromyscus*. *Canadian Journal of Zoology*, **55**, 862–71.
Fairbairn, D.J. (1977b). The spring decline in deer mice: death or dispersal? *Canadian Journal of Zoology*, **55**, 84–92.
Fairbairn, D.J. (1978a). Dispersal in deer mice, *Peromyscus maniculatus*: Proximal causes and the effects on fitness. *Oecologia*, **32**, 171–93.

Fairbairn, D.J. (1978b). Behavior of dispersing deer mice (*Peromyscus maniculatus*). *Behavioral Ecology and Sociobiology*, **3**, 265–82.
Farentinos, R.C. (1972). Social dominance and mating activity in the tassel-eared squirrel (*Sciurus alberti ferreus*). Animal Behaviour, **20**, 316–26.
Fenchel, T. (1974). Intrinsic rate of natural increase: the relationship with body size. *Oecologia*, **14**, 317–26.
Finerty, J.P. (1980). *The Population Ecology of Cycles in Small Mammals*. Yale University Press, New Haven.
Fink, G. (1979). Neuroendocrine control of gonadotrophin secretion. *British Medical Bulletin*, **35**, 155–60.
Fitter, R. (1967). Conservation by captive breeding: a general survey. *Oryx*, **9**, 87–96.
Fitzgerald, B.M. (1977). Weasel predation on a cyclic population of the montane vole (*Microtus montanus*) in California. *Journal of Animal Ecology*, **46**, 367–97.
Fleming, T.H. (1971). Artibeus jamaicensis: delayed embryonic development in a neotropical bat. *Science, New York*, **171**, 402–4.
Fleming, T.H. (1975). The role of small mammals in tropical ecosystems. In *Small Mammals: Their Productivity and Population Dynamics, IBP* 5, pp. 269–98. F. Golley, K. Petrusewicz and L. Ryszkowski (eds). Cambridge University Press, Cambridge.
Fleming, T.H. (1979). Life-history strategies. In *Ecology of Small Mammals*, pp. 1–61. D.M. Stoddart (ed.). Chapman and Hall, London.
Fleming, T.H. and Rauscher, R.J. (1978). On the evolution of litter size in *Peromyscus leucopus*. *Evolution*, **32**, 45–55.
Fletcher, I.C. and Lindsay, D.R. (1968). Sensory involvement in the mating behaviour of domestic sheep. *Animal Behaviour*, **16**, 410–14.
Flint, A.P.F., Renfree, M.B. and Weir, B.J. (eds) (1981). Embryonic Diapause in Mammals. *Journal of Reproduction and Fertility, Supplement* No. 29.
Flowerdew, J.R. (1972). The effect of supplementary food on a population of wood mice (*Apodemus sylvaticus*). *Journal of Animal Ecology*, **41**, 553–66.
Flowerdew, J.R. (1973). The effect of natural and artifical changes in food supply on breeding in woodland mice and voles. *Journal of Reproduction and Fertility, Supplement*, **19**, 259–69.
Flowerdew, J.R (1974). Field and laboratory experiments on the social behaviour and population dynamics of the wood mouse (*Apodemus sylvaticus*). *Journal of Animal Ecology*, **43**, 499–511.
Flowerdew, J.R. (1976). Chapter 4. Ecological methods. *Mammal Review*, **6**, 123–60.
Flowerdew, J.R. Hall, S.J.G. and Brown, J.C. (1977). Small rodents, their habitats, and the effects of flooding at Wicken Fen, Cambridgeshire. *Journal of Zoology, London*, **182**, 323–42.
Flowerdew, J.R. (1985). The population dynamics of wood mice and yellow-necked mice. *Symposia of The Zoological Society of London*, **55**, 315–38.
Flux, J.E.C. (1965). Timing of the breeding season in the hare *Lepus europaeus* (Pallas) and rabbit *Oryctolagus cuniculus* (L.). *Mammalia*, **29**, 557–62.
Follett, B.K. (1973). Circadian rhythms and photoperiodic time measurement in birds. *Journal of Reproduction and Fertility, Supplement*, **19**, 5–18.
Follett, B.K. (1978). Photoperiodism and seasonal breeding in birds and mammals. In *Control of Ovulation*, pp. 267–93. D.B. Crighton, G.R. Foxcroft, N.B. Haynes and G.E. Lamming (eds). Butterworth, London.
Frazer, J.F.D. and Huggett, A. St.G. (1973). Specific foetal growth rates of cetaceans. *Journal of Zoology, London*, **169**, 111–26.
Freeland, W.J. (1974). Vole cycles: another hypothesis. *American Naturalist*, **108**, 238–45.

French, N.R., Stoddart, D.M. and Bobek, B. (1975). Patterns of demography in small mammal populations. In *Small Mammals: Their Productivity and Population Dynamics, IBP* 5, pp. 73–102. F.B. Golley, K. Petrusewicz and L. Ryszkowski (eds). Cambridge University Press, Cambridge.

Fretwell, S. (1969). The adjustment of birth rate to mortality in birds. *Ibis*, **111**, 624–7.

Frye, B.J. (1967) *Hormonal Control in Vertebrates*. Macmillan, New York.

Gadgil, M. and Bossert, W.H. (1970). Life history consequences of natural selection. *American Naturalist*, **104**, 1–24.

Gaines, M.S. and Krebs, C.J. (1971). Genetic changes in fluctuating vole populations. *Evolution*, **25**, 702–23.

Gaines, M.S. and McClenaghan, L. (1980). Dispersal in small mammals. *Annual Review of Ecology and Systematics*, **11** 163–96.

Gaines, M.S., Vivas, A.M. and Baker, C.L. (1979). An experimental analysis of dispersal in fluctuating vole populations: demographic parameters. *Ecology*, **60**, 814–28.

Gambell, R. (1973). Some effects of exploitation on reproduction in whales. *Journal of Reproduction and Fertility, Supplement*, **19**, 531–51.

Gambell, R. (1976a). World whale stocks. *Mammal Review*, **6**, 41–53.

Gambell, R. (1976b). Population biology and management of whales. *Applied Biology*, **1**, 247–343.

Geist, V. (1971). *Mountain Sheep. A Study in Behavior and Evolution*. University of Chicago Press, Chicago.

Geist, V. (1974). On the relationship of social evolution and ecology in ungulates. *American Zoologist*, **14**, 205–20.

Geller, M.D. and Christian, J.J. (1982). Population dynamics, adreno-cortical function and pathology in *Microtus pennsylvanicus*. *Journal of Mammalogy*, **63**, 85–95.

Gibb, J.A. (1981) Limits to population density in the rabbit. In *Proceedings of the World Lagomorph Conference*, pp. 654–63. K. Myers and C.D. MacInnes (eds). University of Guelph.

Gibson, R.M. and Guinness, F.E. (1980). Differential reproduction among red deer (*Cervus elaphus*) stags on Rhum. *Journal of Animal Ecology*, **49**, 199–208.

Gilmore, D. and Cook, B. (eds) (1981). *Environmental Factors in Mammalian Reproduction*. Macmillan, London.

Goldfoot, D.A. Kravetz, M.A., Goy, R.W. and Freeman, S.K. (1976) Lack of effect of vaginal lavages and aliphatic acids on ejaculatory responses in rhesus monkey. Behavioral and chemical analyses. *Hormones and Behavior*, **7**, 1–27.

Golley, F.B., Ryszkowski, L. and Sokur, J.T. (1975). The role of small mammals in temperate forests, grasslands and cultivated fields. In *Small Mammals: Their Productivity and Population Dynamics, IBP* 5, pp. 223–41. F.B. Golley, K. Petrusewicz and L. Ryszkowski (eds). Cambridge University Press, Cambridge.

Goodwin, M., Gooding, K.M. and Regnier, F. (1979). Sex pheromone in the dog. *Science, New York*, **203**, 559–61.

Gosczynski, J. (1977). Connections between predatory birds and mammals and their prey. *Acta Theriologica*, **22**, 399–430.

Gosling, L.M. (1969). Parturition and related behaviour in Coke's Hartebeest, *Alcelaphus buselaphus cokei* Günther. *Journal of Reproduction and Fertility, Supplement*, **6**, 265–86.

Gosling, L.M. (1981). Climatic determinants of spring littering by feral coypus, *Myocastor coypus*. *Journal of Zoology, London*, **195**, 281–8.

Graham, C.F. (1973). Functional microanatomy of the primate uterine cervix. In *Handbook of Physiology Section 7, Vol II part 2*, pp. 1–24. R.O. Greep, and E.B. Astwood (eds). American Physiological Society, Washington.

Graham, J.M. and Desjardins, L. (1980). Classical conditioning: Induction of luteinizing hormone and testosterone secretion in anticipation of sexual activity. *Science, New York*, **210**, 1039–41.

Gregory, E., Engel, K. and Pfaff, D. (1975). Male hamster preference for odors of female hamster vaginal discharges: Studies of experimental and hormonal determinants. *Journal of Comparative Physiology and Psychology*, **89**, 442–6.

Grocock, C.A. (1979). Testis development in the vole *Microtus agrestis* subjected to long or short photoperiods from birth. *Journal of Reproduction and Fertility*, **55**, 423–7.

Grocock, C.A. (1980). Effects of age on photo-induced testicular regression, recrudesence, and refractoryness in the short-tailed field vole *Microtus agrestis. Biology of Reproduction*, **23**, 15–20.

Grocock, C.A. (1981). Effect of different photoperiods on testis weight changes in the vole, *Microtus agrestis. Journal of Reproduction and Fertility*, **68**, 25–32.

Grocock, C.A. and Clarke, J.R. (1974). Photoperiodic control of testis activity in the vole, *Microtus agrestis. Journal of Reproduction and Fertility*, **39**, 337–47.

Grubb, P. (1974). Population dynamics of the Soay sheep. In *Island Survivors: The Ecology of the Soay Sheep of St. Kilda*, pp. 242–72. P.A. Jewell, C. Milner and J. Morton Boyd (eds). The Athlone Press, London.

Grubb, P. and Jewell, P.A. (1973). The rut and the occurrence of oestrus in the Soay sheep of St. Kilda. *Journal of Reproduction and Fertility, Supplement*, **19**, 491–502.

Gurnell, J. (1978). Seasonal changes in numbers and male behavioural interaction in a population of wood mice, *Apodemus sylvaticus. Journal of Animal Ecology*, **47**, 741–55.

Gustafson, A.W. (1979). Male reproductive patterns in hibernating bats. *Journal of Reproduction and Fertility*, **56**, 317–31.

Hafez, E.S.E. (1973). Anatomy and physiology of the mammalian uterotubal junction. In *Handbook of Physiology, Section 7 Endocrinology, Vol. II. part 2.*, pp. 87–95. R.O. Greep and E.B. Astwood (eds). American Physiological Society, Washington.

Hafez, E.S.E. (1974). Functional anatomy of female reproduction. In *Reproduction in Farm Animals*, 3rd edition, pp. 24–53. E.S.E. Hafez (ed.). Lea and Febiger, Philadelphia.

Hairston, N.G., Smith, F.E. and Slobodkin, L.B. (1960). Community structure, population control, and competition. *American Naturalist*, **94**, 421–5.

Hansel, W. and Echternkamp, S.E. (1972). Control of ovarian function in domestic animals. *American Zoologist*, **12**, 225–43.

Hansen, L.P. and Batzli, G.O. (1979). Influence of supplemental food on local populations of *Peromyscus leucopus. Journal of Mammalogy*, **60**, 335–42.

Harison, R.J. (1963). A comparison of factors involved in delayed implantation in badgers and seals in Great Britain. in *Delayed Implantation*, pp. 99–114. A.C. Enders (ed.). University of Chicago Press, Chicago.

Hassell, M.P. (1976). Arthropod predator-prey systems. In *Theoretical Ecology*, pp. 71–93. R.M. May (ed.). Blackwell Scientific Publications, Oxford.

Healey, M.H. (1967). Aggression and self-regulation of population size in deer mice. *Ecology*, **48**, 377–92.

Heap, R.B. and Flint, A.P.F. (1979). Progesterone. In *Reproduction in Mammals. 7. Mechanisms of Hormone Action*, pp. 185–232. C.R. Austin and R.V. Short (eds). Cambridge University Press, Cambridge.

Heap, R.B. and Flint, A.P.F. (1983). Pregnancy. In *Reproduction in Mammals. 3. Hormonal Control of Reproduction*, pp. 153–94. C.R. Austin and R.V. Short (eds). Cambridge University Press, Cambridge.

Heap, R.B., Perry, J.S. and Challis, J.R.G. (1973). Hormonal maintenance of pregnancy. In *Handbook of Physiology, Vol. II part 2*, pp. 217–60. R.O. Greep and E.B. Astwood (eds). American Physiological Society, Washington.
Heller, C.G. and Clermont, Y. (1964). Kinetics of the germinal epithelium in man. *Recent Progress in Hormone Research*, **20**, 545–75.
Henderson, B.A. (1979). Regulation of the size of the breeding population of the European rabbit, *Oryctolagus cuniculus*, by social behaviour. *Journal of Animal Ecology*, **16**, 383–92.
Henderson, B.A. (1981). The role of food quality and social behaviour in the regulation of the size of the breeding population of the European rabbit. In *Proceedings of the World Lagomorph Conference, 1979*, pp. 664–6. K. Myers and C.D. McInnes (eds). University of Guelph.
Henderson, K.M. (1979). Gonadotrophin regulation of ovarian activity. *British Medical Bulletin*, **35**, 161–6.
Hendrichs, H. (1975). Changes in a population of Dikdik, *Madoqua (Rhynchotragus) kerki* (Günther, 1880). *Zeitschrift für Tierpsychology*, **38**, 55–69.
Hendrichs, H. and Hendrichs, U. (1971). *Dikdik und elephanten Ökologie und Soziologie zweier afrikanischer Huftiere*. R. Piper and Co., Munich.
Herbert, J., Stacey, P.M. and Thorpe, D.H. (1978). Recurrent breeding seasons in pinealectomized or optic nerve sectioned ferrets. *Journal of Endocrinology*, **78**, 389–97.
Hestbeck, J.B. (1982). Population regulation of cyclic mammals: the social fence hypothesis. *Oikos*, **69**, 157–63.
Hewer, H.R. (1974). *British Seals*. Taplinger Publishing Co., New York.
Hill, D.A. (1984). Population regulation in the mallard (*Anas platyrhynchos*). *Journal of Animal Ecology*, **53**, 191–202.
Hoffman, J.C. (1973). The influence of photoperiods on reproductive functions in female mammals. In *Handbook of Physiology, Section 7, Vol. II part 1*, pp. 57–77. R.O. Greep and E.B. Astwood. American Physiological Society, Washington.
Hoffman, K. Illnerová, H. and Vanécek, J. (1981). Effect of photoperiod and of one minute light at night-time on the pineal rhythm on (sic) N-Acetyltransferase activity in the Djungarian hamster *Phodopus sungorus*. *Biology of Reproduction*, **24**, 551–6.
Holling, C.S. (1959). The components of predation as revealed by a study of small-mammal predation on the European pine sawfly. *The Canadian Entomologist*, **91**, 234–61.
Holling, C.S. (1973). Resiliance and stability of ecological systems. *Annual Review of Ecology and Systematics*, **4**, 1–23.
Holgate, P. (1967). Population survival and life history phenomena. *Journal of Theoretical Biology*, **14**, 1–10.
Hooper, E.T. (1958). The male phallus in mice of the genus *Peromyscus*. *Miscellaneous Publications of the Museum of Zoology, University of Michigan*, **105**, 1–24.
Horncocker, M.G. (1969). Winter territoriality in mountain lions. *Journal of Wildlife Management*, **33**, 457–64.
Horncocker, M.G. (1979). An analysis of mountain lion predation upon mule deer and elk in the Idaho primitive areas. *Wildlife Monographs*, **21**, 39pp.
Horst, C.J. van der, and Gillman, J. (1940). Ovulation and corpus luteum formation in *Elephantulus*. *South African Journal of Medical Science*, **5**, 73–91.
Howard, L.O. and Fiske, W.F. (1911). The importation into the United States of the parasites of the gypsy moth and the brown-tail moth. *Bulletin of the U.S. Bureau of Entomology*, **91**, 1–344.
Hrdy, S.B. (1974). Male-male competition and infanticide among langurs (*Presbytis entellus*) of Abu, Rajasthan. *Folia Primatologica*, **22**, 19–58.

Huffaker, C.B. and Messanger, P.S. (1964). The concept and significance of natural control. In *Biological Control of Insect Pests and Weeds*, pp. 74–117. P. DeBach (ed.). Chapman & Hall, London.
Huggett, A. St. G. and Widdas, W.F. (1951). The relationships between mammalian foetal weight and conception age. *Journal of Physiology, London*, **114**, 306–17.
Hunter, R.H.F. (1980). *Physiology and Technology of Reproduction in Female Domestic Animals*. Academic Press, London.
Huxley, J.S. (1938). The present standing of the theory of sexual selection. In *Evolution*, pp. 11–42. G. DeBeer (ed.). Oxford University Press, Oxford.
Hyashi, S. and Kimura, T. (1974). Sex attractant emitted by female mice. *Physiology and Behavior*, **13**, 563–7.
Itô, Y. (1978). *Comparative Ecology*. Cambridge University Press, Cambridge.
Izard, M.K. and Vandenbergh, J.G. (1982). Priming pheromones from oestrous cows increase synchronisation of oestrus in dairy heifers after PGF-2α injection. *Journal of Reproduction and Fertility*, **66**, 189–96.
Jarman, P.J. (1974). The social organisation of antelope in relation to their ecology. *Behaviour*, **48**, 215–67.
Jenni, D.A. (1974). Evolution of polyandry in birds. *American Zoologist*, **14**, 129–44.
Jensen, E.V. (1979). The oestrogens. In *Reproduction in Mammals. 7. Mechanisms of Hormone Action*, pp. 157–184. C.R. Austin and R.V. Short (eds).
Jensen, T.S. (1982). Seed production and outbreaks of non-cyclic rodent populations in deciduous forests. *Oecologia*, **54**, 184–92.
Jewell, P.A. (1966). Breeding season and recruitment in some British mammals confined on small islands. *Symposia of the Zoological Society of London*, **15**, 89–116.
Jewell, P.A. (1977). The evolution of mating systems in mammals. In *Reproduction and Evolution*, pp.23–32. J.H. Calaby and C.H. Tyndale-Biscoe (eds). Australian Academy of Science, Canberra.
Johns, M.A., Feder, H.H., Komisaruk, B.R. and Mayer, A.D. (1978). Urine induces reflex ovulation in anovulatory rats: a vomeronasal effect. *Nature, London*, **272**, 446–7.
Johnson, R.P. (1973). Scent marking in mammals. *Animal Behaviour*, **21**, 521–35.
Johnston, P.G. and Zucker, I. (1980). Photoperiodic regulation of the testis of adult white-footed mice (*Peromyscus leucopus*). *Biology of Reproduction*, **23**, 859–66.
Josso, N. (1974). *In vitro* synthesis of Mullerian-inhibiting hormone by seminiferous tubules isolated from calf fetal testis. *Endocrinology*, **93**, 829–34.
Josso, N., Picard, J.Y. and Tran, D. (1977). The anti-Müllerian hormone. *Recent Progress in Hormone Research*, **33**, 117–160.
Jost, A. (1965). Gonadal hormones in sex differentiation of the mammalian fetus. In *Organogenesis*, pp. 611–52. L.R.L. Dehaan and H. Ursprung (eds). Holt, Rinehart and Winston, New York.
Jost, A. (1979). Basic sexual trends in the development of vertebrates. In *Sex, Hormones and Behaviour, Ciba Foundation Symposium*, **62**, 5–18.
Jost, A. Vigier, B., Prépin, J. and Perchellet, J.P. (1973). Studies on sex differentiation in mammals. *Recent Progress in Hormone Research*, **29**, 1–42.
Kaczmarski, F. (1966). Bioenergetics of pregnancy and lactation in the bank vole. *Acta theriologica*, **11**, 409–17.
Kalela, O. (1962). On the fluctuations in the numbers of arctic and boreal small rodents as a problem of production ecology. *Annales Academiae Scientarum Fennicae Series A IV. Biology*, **66**, 1–38.
Karsch, F.J. (1984). The hypothalamus and anterior pituitary gland. In *Reproduction in Mammals. 3. Hormonal Control of Reproduction*, 2nd edition, C.R. Austin and R.V. Short (eds). Cambridge University Press, Cambridge.

Karsch, F.J. and Foster, D.L. (1981). Environmental control of seasonal breeding: a common final mechanism governing seasonal breeding and sexual maturation. In *Environmental Factors in Mammal Reproduction*, pp. 30–53. D. Gilmore and B. Cook (eds). Macmillan, London.

Karsch, F.J., Goodman, R.L. and Legan, S.J. (1980). Feedback basis of seasonal breeding: test of an hypothesis. *Journal of Reproduction and Fertility*, **58**, 521–35.

Kear, J. (1977). The problems of breeding endangered species in captivity. In *International Zoo Yearbook 17*, pp. 5–13. P.J.S. Olney, (ed.). Zoological Society of London, London.

Keith, L.B. (1974). Some features of population dynamics in mammals. *International Congress of Game Biologists, Sweden*, **11**, 17–58.

Keith, L.B. (1983). Role of food in hare population cycles. *Oikos*, **40**, 385–95.

Keith, L.B., Rongstadt, O.J. and Meslow, E.C. (1966). Regional differences in reproductive traits of the snowshoe hare. *Canadian Journal of Zoology*, **44**, 953–61.

Keith, L.B., Todd, A.W., Brand, C.J., Adamcik, R.S. and Rusch, D.H. (1977). An analysis of predation during a cyclic fluctuation of snowshoe hares. *International Congress of Game Biologists, Atlanta, Georgia*, **13**, 151–75. Wildlife Society, Washington.

Kenny, A. McM. and Dewsbury, D.A. (1977). Effect of limited mating on the corpora lutea in montane voles, *Microtus montanus*. *Journal of Reproduction and Fertility*, **49**, 363–4.

King, C.M. (1977). Stoat *Mustela erminea*. In *The Handbook of British Mammals*, pp. 331–8. G.B. Corbet and H.N. Southern (eds). Blackwell Scientific Publications, Oxford.

Kleiman, D.G. (1968). Reproduction in the Canidae. In *International Zoo Yearbook 8*, pp. 3–8. N. Duplaix-Hall (ed.). Zoological Society of London, London.

Kleiman, D.G. (1977). Monogamy in mammals. *Quarterly Review of Biology*, **52**, 39–69.

Kleiman, D. G. and Eisemberg, J.F. (1973). Comparisons of canid and felid social systems from an evolutionary perspective. *Animal Behaviour*, **21**, 637–59.

Klein, D.C. and Weller, J.E. (1972). Rapid light induced descrease in pineal serotonin N-acetyl l-transferase activity. *Science, New York*, **177**, 532–3.

Klein, D.R. (1981). The problems of overpopulation of deer in North America. In *Problems in Management of Locally Abundant Wild Mammals*, pp. 119–27. P.A. Jewell and S. Holt (eds). Academic Press, New York.

Klingel, H. (1968). Soziale Organisation und verhaltenweisen von Hartmann- und Bergzebras (*Equus zebra hartmannae* und *E. z. zebra*). *Zeitschrift Für Tierpsychologie*, **25**, 76–88.

Klingel, H. (1969). The social organisation and population ecology of the plains zebra (*Equus quagga*). *Zoologica Africana*, **4**, 249–63.

Klomp. H. (1962). The influence of climate and weather on the mean density level, the fluctuations and the regulation of animal populations. *Archives Neerlandaises de Zoologie*, **15**, 68–109.

Klomp. H. (1970). Clutch size in birds. *Ardea*, **58**, 1–121.

Klopfer, P.H. and Gambale, J. (1966). Maternal imprinting in goats: the role of chemical senses. *Zeirschrift für Tierpsychologie*, **23**, 588–92.

Knight, T.W. and Lynch, P.R. (1980). Source of ram pheromones that stimulate ovulation in the ewe. *Animal Reproductive Science*, **3**, 133–6.

Knobil, E. (1980). The neuroendocrine control of the menstrual cycle. *Recent Progress in Hormone Research*, **36**, 53–88.

Knudsen, B. (1962). Growth and reproduction of house mice at three different temperatures. *Oikos*, **13**, 1–14.

Krebs, C.J. (1964). The lemming cycle at Baker Lake, Northwest Territories, during 1959–1962. *Arctic Institute of North America Technical paper No. 15.*

Krebs, C.J. (1970). *Microtus* population biology: behavioural changes associated with the population cycle in *Microtus ochrogaster* and *M. pennsylvanicus. Ecology*, **51**, 34–52.

Krebs, C.J. (1978a). A review of the Chitty hypothesis of population regulation. *Canadian Journal of Zoology*, **56**, 2463–80.

Krebs, C.J. (1978b). Dispersal and cyclic changes in populations of small rodents. In *Aggression, Dominance and Individual Spacing*, pp. 49–60. L. Krames, P. Pliner and T. Alloway (eds). Plenum Publishing Corporation, New York.

Krebs, C.J. (1985). *Ecology. The Experimental Analysis of Distribution and Abundance* 3rd edition. Harper and Row, New York.

Krebs, C.J. and DeLong, K.T. (1965). A *Microtus* population with supplemental food. Journal of Mammalogy, **46**, 566–73.

Krebs, C.J. and Myers, J.H. (1974). Population cycles in small mammals. *Advances in Ecological Research*, **8**, 268–399.

Krebs, C.J., Keller, B.L. and Tamarin, R.H. (1969). *Microtus* population biology: demographic changes in fluctuating populations of *M. ochrogaster* and *M. pennsylvanicus* in southern Indiana. *Ecology*, **50**, 587–607.

Krebs, C.J., Wingate, J., LeDuc, J., Redfield, J.A., Taitt, M. and Hilborn, R. (1976). *Microtus* population biology: dispersal in fluctuating populations of *M. townsendii. Canadian Journal of Zoology*, **54**, 79–95.

Krebs, J.R. and Davies, N.B. (eds) (1978). *Behavioural Ecology, an Evolutionary Approach*. Blackwell Scientific Publications, Oxford.

Kretser, D.M. de, Bremner, W.J., Burger, H.G., Hudson, B., Irby, D.C., Kerr, J.B. and Lee, V.W.K. (1977). The interaction between the hypothalamo-hypophyseal system and the testis in mammalian species. In *Reproduction and Evolution*, pp. 111–9. J.H. Calaby and C.H. Tyndale-Biscoe (eds). Australian Academy of Science, Canberra.

Krzanowska, H. (1974). The passage of abnormal spermatozoa through the utero-tubal junction of the mouse. *Journal of Reproduction and Fertility*, **38**, 81–90.

Kummer, H. (1968). *Social Organization of Hamadryas Baboons: A Field Study*. University of Chicago Press, Chicago; (Bibliotheca Primatologica No. 6), S. Kager, Basel.

Lack, D. (1947–48). The significance of clutch size. *Ibis*, **89**, 302–52; **90**, 25–45.

Lack, D. (1954). *The Natural Regulation of Animal Numbers*. Clarendon Press, Oxford.

Lack, D. (1966). *Population Studies of Birds*. Oxford University Press, Oxford.

Ladewig, J. and Hart, B.L. (1980). Flehmen and vomeronasal function in male goats. *Physiology and Behaviour*, **24**, 1067–71.

Laine, K. and Henttonen, H. (1983). The role of plant production in microtine cycles in northern Fennoscandia. *Oikos*, **40**, 407–18.

Laws, R.M., Parker, I.S.C. and Johnstone, R.C.B. (1975). *Elephants and Their Habitats*. Oxford University Press, Oxford.

LeBoeuf, B.J. (1972). Sexual behaviour in the northern elephant seal *Mirounga angustirostris. Behaviour*, **41**, 1–26.

LeBoeuf, B.J. (1974). Male-male competition and reproductive success in elephant seals. *American Zoologist*, **14**, 163–76.

LeBoeuf, B.J. and Peterson, R.S. (1969). Social status and mating activity in elephant seals. *Science, New York*, **163**, 91–3.

LeDuc, J. and Krebs, C.J. (1975). Demographic consequences of artificial selection of the LAP locus in voles (*Microtus townsendii*). *Canadian Journal of Zoology*, **53**, 1825–40.

Lee, A.K., Bradley, A.J. and Braithwaite, R.W. (1977). Corticosteroid levels and male mortality in *Antechinus stuartii*. In *Biology and Environment 2. Biology of Marsupials*, pp. 209–20. B. Stonehouse and D. Gilmore (eds). Macmillan, London.
Lee, S. van der and Boot, L.M. (1956). Spontaneous pseudopregnancy in mice I. *Acta Physiologica et Pharmagologica Neerlandica*, **4**, 442–4.
Lee, S. van der and Boot, L.M. (1956). Spontaneous pseudopreganancy in mice II. *Acta Physiologica et Pharmacologica Neerlandica*, **5**, 213–4.
Leidahl, L.C. and Moltz, H. (1975). Emission of maternal pherome in the nulliparous female and failure of emission in the adult male. *Physiology and Behavior*, **14**, 421–4.
Leitch, I. Hytten, F.E. and Billewicz, W.Z. (1959). The maternal and neonatal weights of some mammalia. *Proceedings of the Zoological Society of London*, **133**, 11–28.
Leon, M. and Moltz, H. (1972). The development of the pheromonal bond in the albino rat. *Physiology and Behaviour*, **8**, 683–6.
LeMagen, J. (1952). Les phénoménes olfacto-sexuels chez le rat blanc. *Archives des Sciences Physiologiques*, **6**, 295–332.
Lent, P.C. (1974). Mother-infant relationships in Ungulates . In *The Behaviour of Ungulates and its Relation to Management*, pp. 144–55. V. Geist and F. Walther (eds). IUCN New Series No. 24, IUCN, Morges.
Leslie, P.H. and Ranson, R.M. (1940). The mortality, fertility rate of natural increase of the vole (*Microtus agrestis*) as observed in the laboratory. *Journal of Animal Ecology*, **9**, 27–52.
Lett, P.F. and Benjaminsen, T. (1977). A stochastic model for the management of the Northwestern Atlantic harp seal (*Pagophilus groenlandicus*) population. *Journal of the Fisheries Research Board of Canada*, **34**, 1155–87.
Lett, P.F., Mohn, R.K. and Grey, D.F. (1979). Density-dependent processes and management strategy for the Northwest Atlantic harp seal population. *International Commission for the Northwest Atlantic Fisheries Selected Papers*, **5**, 61–80.
Lidicker, W.Z. Jr. (1962). Emigration as a possible mechanism permitting the regulation of population density below carrying capacity. *American Naturalist*, **96**, 29–33.
Lidicker, W.Z. Jr. (1973). Regulation of numbers in an island population of the California vole: A problem in community dynamics. *Ecological Monographs*, **43**, 271–302.
Lidicker, W.Z. Jr. (1975). The role of dispersal in the demography of small mammals. In *Small Mammals: Their Productivity and Population Dynamics, IBP 5*, pp. 103–34. F.B. Golley, K. Petrusewicz and L. Ryszkowski (eds). Cambridge University Press, Cambridge.
Lidicker, W.Z. Jr. (1978). Regulation of numbers in small mammal populations – Historical reflections and a synthesis. In *Population of Small Mammals Under Natural Conditions*, pp. 122–41. D.P. Snyder (ed.). Special Publication Series. Vol. 5, Pymatuning Laboratory of Ecology, University of Pittsburgh.
Lincoln, G.A. (1978). Induction of testicular growth and sexual activity in rams by 'skeleton' short-day photoperiod. *Journal of Reproduction and Fertility*, **52**, 179–81.
Lincoln, G.A. (1979). Pituitary control of testicular activity. *British Medical Bulletin*, **75**, 167–72.
Lincoln, G.A. and Davidson, W. (1977). The relationship between sexual and aggressive behaviour and pituitary and testicular activity during the seasonal sexual cycle of rams, and the influence of photoperiod. *Journal of Reproduction and Fertility*, **49**, 267–76.

Lincoln, G.A. and Guinness, F.E. (1973). The sexual significance of the rut in red deer. *Journal of Reproduction and Fertility, Supplement*, **19**, 475–89.

Lincoln, G.A. and Short, R.V. (1980). Seasonal breeding: Nature's contraceptive. *Recent Progress in Hormone Research*, **36**, 1–52.

Lindsay, D.R. (1965). The importance of olfactory stimuli in the mating behaviour of the ram. *Animal Behaviour*, **13**, 75–8.

Lloyd, G. (1977). Fox *Vulpes vulpes*. In *The Handbook of British Mammals*, pp. 311–20. G.B. Corbet and H.N. Southern. (eds). Blackwell Scientific Publications, Oxford.

Lofts, B. (1970). *Animal Photoperiodism*. Edward Arnold, London.

Lord, R.D. Jr. (1960). Litter size and latitude in North American mammals. *American Midland Naturalist*, **64**, 488–99.

Lowe, V.P.W. (1977). Red deer *Cervus elaphus*. In *The Handbook of British Mammals*, pp. 411–23. G.B. Corbet and H.N. Southern (eds). Blackwell Scientific Publications, Oxford.

Loy, J. (1970). Peri-menstrual sexual behaviour among rhesus monkeys. *Folia Primatologica*, **13**, 286–97.

MacArthur, R.H. and Wilson, E.O. (1967). *Theory of Island Biogeography*. Princeton University Press, Princeton.

McCarty, R. and Southwick, C.H. (1977). Cross-species fostering: Effects on the olfactory preference of *Onychomys torridus* and *Peromyscus leucopus*. *Behavioral Biology*, **19**, 255–60.

McClure, P.A. (1981). Sex-biased litter reduction in food-restricted wood rats (*Neotoma floridana*). *Science, New York*, **211**, 1058–60.

McClure, T.J. (1962). Infertility in female rodents caused by temporary inanition at or about the time of implantation. *Journal of Reproduction and Fertility*, **4**, 241.

Macdonald, D.W. (1977). On food preference in the red fox. *Mammal Review*, **7**, 1–23.

McGill, T.E. and Loughlin, R.C. (1970). Ejaculatory reflex and luteal activity induction in *Mus musculus*. Journal of *Reproduction and Fertility*, **21**, 215–20.

McKeown, T., Marshall, T. and Record, R.G. (1976). Influences on fetal growth. *Journal of Reproduction and Fertility*, **47**, 167–81.

MacKinnon, J. (1974a). *In Search of the Red Ape*, Collins, London.

MacKinnon, J. (1974b). The behaviour and ecology of wild orang-utan (*Pongo pygmaeus*). *Animal Behaviour*, **22**, 3–74.

MacKinnon, J. (1978). *The Ape Within Us*. Collins, London.

MacKinnon, J. (1979). Reproductive behaviour in wild orang-utan populations. In *The Great Apes*, D.A. Hamburg and E.R. McCown (eds). Benjamin/Cummings, Menlo Park, California.

MacKinnon, K. (1978). Competition between red and grey squirrels. *Mammal Review*, **8**, 185–90.

Mallory, F.F. and Brooks, R.J. (1978). Infanticide and other reproductive strategies in the collared lemming, *Dicrostonyx groenlandicus*. *Nature, London*, **273**, 144–6.

Mallory, F.F. and Clulow, F.V. (1977). Evidence of pregnancy failure in the wild meadow vole, *Microtus pennsylvanicus*. *Canadian Journal of Zoology*, **55**, 1–17.

Mann, T. (1964). *The Biochemistry of Semen and of the Male Reproductive Tract*. Methuen, London.

Mann, T. (1974). Secretory function of the prostate, seminal vesicle and the male accessory organs of reproduction. *Journal of Reproduction and Fertility*, **37**, 179–88.

Marchlewska-Koj, A. (1984). Pheromones and mammalian reproduction. *Oxford Reviews of Reproductive Biology*, **6**, 266–302.

Marsden, H.M. and Bronson, F.H. (1964). Estrous synchrony in mice: alternation by exposure to male urine. *Science, New York*, **144**, 3625.
Marsden, H.M. and Conaway, C.H. (1963). Behavior and the reproductive cycle in the cottontail. *Journal of Wildlife Management*, **27**, 161–70.
Marshall, A.J. and Corbet, P.S. (1959). The breeding biology of equatorial vertebrates: Reproduction of the bat *Chaerophon hindei* Thomas at 0° 26′N. *Proceedings of the Zoological Society of London*, **132**, 607–16.
Marshall, F.H.A. (1937). On the changeover in oestrous cycles in animals after transference across the equator, with further observations on the incidence of the breeding seasons and the factors controlling sexual periodicity *Proceedings of the Royal Society (B)*, **122**, 413–28.
Marshall, F.H.A. (1942). Exteroceptive factors in sexual periodicity. *Biological Reviews*, **17**, 68–89.
Marshall, W.A. (1978). The relationship of puberty to other maturity indicators and body composition in man. *Journal of Reproduction and Fertility*, **52**, 437–43.
Maruniak, J.A., Desjardins, C. and Bronson, F.H. (1975). Adaptations for urinary marking in rodents: prepuce length and morphology. *Journal of Reproduction and Fertility*, **44**, 567–70.
Matthews, L. Harrison, (1935). The oestrous cycle and intersexuality in the female mole (*Talpa europaea* Linn.). *Proceedings of the Zoological Society of London*, **1935**, 347–83.
Mattner, P.E. (1966). Formation and retention of the spermatozoan reservoir in the cervix of the ruminant. *Nature*, **312**, 1479–80.
Mauleon, P. and Mariana, J.C. (1977). Oogenesis and folliculogenesis. In *Reproduction in Domestic Animals*, pp.175–202. H.H. Cole and P.T. Cupps (eds). Academic Press, London.
May, R.M. (ed.) (1976). *Theoretical Ecology*, Blackwell Scientific Publications, Oxford.
May, R.M. (1976). Harvesting whale and fish populations. *Nature*, **263**, 91–2.
May, R.M. (ed.) (1981). *Theoretical Ecology*, 2nd edition, Blackwell Scientific Publications, Oxford.
May, R.M. and Anderson, R.M. (1979). Population biology of infectious diseases: Part II. *Nature*, **280**, 455–61.
Mayr, E. (1939). The sex ratio in wild birds. *American Naturalist*, **73**, 156–79.
Mech, L.D. (1970). *The Wolf*. Doubleday, New York.
Mech, L.D. (1977). Wolf-pack buffer zones as prey reservoirs. *Science, New York*, **198**, 320–21.
Michael, R.P. and Keverne, E.B. (1970). Primate sex pheromones of vaginal origin. *Nature*, **225**, 84–5.
Michael, R.P., Keverne, E.B. and Bonsall, R.W. (1971). Pheromones: isolation of male sex attractants from a female primate. *Science, New York*, **172**, 964–6.
Millar, J.S. (1977). Adaptive features of mammalian reproduction. *Evolution*, **31**, 370–86.
Milligan, S.R. (1974). Social environment and ovulation in the vole, *Microtus agrestis*. *Journal of Reproduction and Fertility*, **41**, 35–47.
Milligan, S.R. (1975). Mating, ovulation and corpus luteum function in the vole, *Microtus agrestis*. *Journal of Reproduction and Fertility*, **42**, 35–44.
Milligan, S.R. (1979). The copulatory pattern of the bank vole (*Clethrionomys glareolus*) and speculation on the role of penile spines. *Journal of Zoology, London*, **188**, 279–83.
Mitchell, B. (1973). The reproductive performance of wild Scottish red deer *Cervus elaphus*. *Journal of Reproduction and Fertility*, **19**, 271–85.
Monder, H., Lee, C-T., Donovick, P.J. and Burright, R.G. (1978). Male mouse urine

extract effects on pheromonally mediated reproductive functions of female mice. *Physiology and Behavior*, **20**, 447–52.

Monesi, V. (1977). Spermatogenesis and spermatozoa. In *Reproduction in Mammals. 1. Germ Cells and Fertilization*, pp. 46–84. C.R. Austin and R.V. Short (eds). Cambridge University Press, Cambridge.

Montgomery, W.I. (1985). Interspecific competition and comparative ecology of *Apodemus* species. In *Case Studies in Population Biology*, pp. 126–87. L.M. Cook (ed.). Manchester University Press, Manchester.

Morton, D.B. and Glover, T.D. (1974). Sperm transport in the female rabbit: the role of the cervix. *Journal of Reproduction and Fertility*, **38**, 131–8.

Moss, R., Watson, A. and Ollason, J. (1982). *Animal Population Dynamics*. Chapman and Hall, London.

Mountford, M.D. (1973). The significance of clutch size. In *The Mathematical Theory of the Dynamics of Biological Populations*, pp. 315–23. M.S. Bartlett and R.W. Hiorns (eds). Academic Press, London.

Muller-Schwarze, D. (1979). Flehmen in the context of mammalian communication. In *Chemical Ecology: Odour Communication in Animals*, pp. 85–96. F.J. Ritter (ed.). Elsevier, North Holland.

Muller-Schwarze, D. and Silverstein, R.M. (eds) (1980). *Chemical Signals: Vertebrates and Aquatic Vertebrates*. Plenum Press, New York.

Murdoch, W.W. (1970). Population regulation and population inertia. *Ecology*, **51**, 497–502.

Murdoch, W.W. and Oaten, A. (1975). Predation and population stability. *Advances in Ecological Research*, **9**, 1–131.

Murphy, G.I. (1968). Patterns in life history and the environment. *American Naturalist*, **102**, 391–403.

Mutere, F.A. (1967). Delayed implantation in an equatorial fruit bat. *Nature, London*, **207**, 780.

Myers, J. (1974). Genetic and social structure of feral house mouse populations on Grizzly Island, California. *Ecology*, **55**, 747–59.

Myers, J.H. and Krebs, C.J. (1971). Genetic, behavioural and reproductive attributes of dispersing field voles, *Microtus pennsylvanicus* and *Microtus ochrogaster* *Ecological Monographs*, **41**, 53–78.

Myers, K. and Poole, W.E. (1962). A study of the biology of the wild rabbit *Oryctolagus cuniculus* (L.) in confined populations. III. Reproduction. *Australian Journal of Zoology*, **10**, 225–67.

Mykytowycz, R. and Fullager, P.J. (1973). Effect of social environment on reproduction in the rabbit. *Oryctolagus cuniculus* (L.). *Journal of Reproduction and Fertility Supplement*, **19**, 503–22.

Myllimaki, A. (1977). Demographic mechanisms in fluctuating populations of the field vole *Microtus agrestis*. *Oikos*, **29**, 468–93.

Nalbandov, A.V. (1964). *Reproductive Physiology*, 2nd edition. W.H. Freeman & Co., San Francisco.

Negus, N.C., Berger, P.J. and Forslund, L.G. (1977). Reproductive strategy of *Microtus montanus*. Journal of Mammalogy, **58**, 347–53.

Newsome, A.E. (1969). A population study of house-mice temporarily inhabiting a South Australian wheatfield. *Journal of Animal Ecology*, **38**, 341–59.

Newsome, A.E. (1970). An experimental attempt to produce a mouse plague. *Journal of Animal Ecology*, **39**, 299–311.

Newsome, A.E. (1973). Cellular degeneration in the testis of red kangaroos during hot weather and drought in central Australia. *Journal of Reproduction and Fertility, Supplement*, **19**, 191–201.

Newson, R.M. (1966). Reproduction of the feral coypu (*Myocastor coypus*).

Symposia of the Zoological Society of London, **15**, 323–4.
Nicholson, A.J. (1933). The balance of animal populations. *Journal of Animal Ecology Supplement*, **2**, 132–78.
Nicholson, A.J. (1954). An outline of the dynamics of animal populations. *Australian Journal of Zoology*, **2**, 9–65.
O'Connor, F.B. (1974). The ecological basis for conservation. In *Conservation in Practice*, pp. 87–98. A. Warren and F.B. Goldsmith, (eds). John Wiley & Son Ltd., London.
Ohno, S. (1977). Homology of X-linked genes in mammals and evolution of sex-determining mechanisms. In *Reproduction and Evolution*, pp. 49–53. J.H. Calaby and C.H. Tyndale-Biscoe (eds). Australian Academy of Science.
Oldemeyer, J.L., Franzman, A.W., Brundage, A.L. Arneson, P.D. and Flynn, A. (1977). Browse quality and the Kenai moose population. *Journal of Wildlife Management*, **41**, 533–42.
Orians, G.H. (1969). On the evolution of mating systems in birds and mammals. *American Naturalist*, **103**, 589–603. The University of Chicago Press.
Owen-Smith, N. (1977). On territoriality in ungulates and an evolutionary model. *Quarterly Review of Biology*, **52**, 1–38.
Oxberry, B.A. (1979). Female reproductive patterns in hibernating bats. *Journal of Reproduction and Fertility*, **56**, 359–67.
Packard, J.M. and Mech, L.D. (1980). Population regulation in wolves. In *Biosocial Mechanisms of Population Regulation*, pp. 135–49. M.N. Cohen, R.S. Malpass and H.G. Klein (eds). Yale University Press, New Haven.
Parker, G.A. (1970). Sperm competition and its evolutionary consequences in the insects. *Biological Reviews*, **45**, 525–68.
Pease, J.L., Vowles, R.H. and Keith, L.B. (1979). Interaction of snowshoe hares and woody vegetation. *Journal of Wildlife Management*, **43**, 43–60.
Pearson, O.P. (1944). Reproduction in the shrew (*Blarina brevicauda*, Say). *American Journal of Anatomy*, **75**, 39–93.
Pearson, O.P. (1963). History of two local outbreaks of feral house mice. *Ecology*, **44**, 540–9.
Pearson, O.P. (1966). The prey of carnivores during one cycle of mouse abundance. *Journal of Animal Ecology*, **35**, 217–33.
Pearson, O.P. (1971). Additional measurements of the impact of carnivores on California voles (*Microtus californicus*). *Journal of Mammalogy*, **52**, 41–9.
Pearson, O.P. and Enders, R.K. (1944). Duration of pregnancy in certain mustelids. *Journal of Experimental Zoology*, **95**, 21–35.
Pederson, T. (1970). Follicle kinetics in the ovary of the cyclic mouse. *Acta Endocrinologica*, **64**, 304–23.
Perey, B., Clermont, Y. and Leblond, C.P. (1961). The wave of the seminiferous epithelium in the rat. *American Journal of Anatomy*, **108**, 47–77.
Perry, J. (1979). Reintroduction hazards. *Oryx*, **15**, 80.
Perry, J.S. (1945). Reproduction of the wild brown rat. (*Rattus norvegicus* Erzleben). *Proceedings of the Zoological Society of London*, **115**, 19–46.
Perry, J.S. (1971). *The Ovarian Cycle of Mammals*. Oliver & Boyd, Edinburgh.
Peters, H., Byskov, A.G., Himelstein-Braw, R. and Faber, M. (1975). Follicular growth: the basic event in the mouse and human ovary. *Journal of Reproduction and Fertility*, **45**, 559–66.
Phillipson, J. (1973). The biological efficiency of protein production by grazing and other land based systems. In *The Biological Efficiency of Protein Production*, pp. 217–35. J.G.W. Jones. Cambridge University Press, Cambridge.
Piacsek, B.E. and Nazian, S.J. (1981). Thermal influences on sexual maturation in the rat. In *Environmental factors in Mammal Reproduction*, pp. 214–31. D.

Gilmore and B. Cook (eds). Macmillan, London.
Pianka, E.R. (1970). On 'r' and 'K' selection. *American Naturalist*, **104**, 592–7.
Pianka, E.R. (1972). 'r' and 'K' or 'b' and 'd' selection. *American Naturalist*, **106**, 581–8. The University of Chicago Press.
Pickard, G. (1977). Changes in hypothalamic luteinizing releasing hormone (LH-RH) in the male golden hamster in response to shortened photoperiods. *Society for Neuroscience* III, **354** (Abstract).
Pielou, E.C. (1974). *Population and Community Ecology*. Gordon & Breach, New York.
Pierrepoint, C.G., Davies, P., Lewis, M.H. and Moffat, D.B. (1975). Examination of the hypothesis that a direct control system exists for the prostate and seminal vesicles. *Journal of Reproduction and Fertility*, **44**, 395–409.
Pimlott, D.H. (1967). Wolf predation and ungulate populations. *American Zoologist*, **7**, 267–78.
Pinder, N.J. and Barkham, J.P. (1978). An assessment of the contribution of captive breeding to the conservation of rare mammals. *Biological Conservation*, **13**, 187–245.
Pitelka, F.A. (1957). Some characteristics of microtine cycles in the Arctic. In *Arctic Biology*, pp. 153–84. H.P. Hansen (ed.). Oregon State University Press, Corvallis.
Pitalka, F.A. and Schultz, A.M. (1964). The nutrient-recovery hypothesis for arctic microtine rodents. In *Grazing in Terrestrial and Marine Environments*, pp. 58–68. D. Crisp. (ed.). Blackwell Scientific Publications, Oxford.
Platt, C. (1978). Conservation versus commercial exploitation of seals with special reference to the Harp seal. *Mammal Review*, **8**, 15–18.
Podoler, H. and Rogers, D. (1975). A new method for the identification of key factors from life-table data. *Journal of Animal Ecology*, **44**, 85–114.
Porter, R.H. and Doane, H.M. (1976). Maternal pheromone in the spiny mouse (*Acomys caharinus*). *Physiology and Behavior*, **16**, 75–8.
Porter, R.H., Deni, R. and Doane, H.M. (1977). Responses of *Acomys cahirinus* pups to chemical cues produces by a foster species. *Behavioral Biology*, **20**, 244–51.
Powers, J.B. and Winans, S.S. (1975). Vomeronasal organ: Critical role in mediating sexual behavior of the male hamster. *Science, New York*, **187**, 961–3.
Putman, R.J. and Wratten, S.D. (1984). *Principles of Ecology*. Croom Helm, London.
Racey, P.A. (1973). Environmental factors affecting the length of gestation in heterothermic bats. *Journal of Reproduction and Fertility, Supplement*, **19**, 175–89.
Racey, P.A. (1979). The prolonged storage and survival of spermatozoa in Chiroptera. *Journal of Reproduction and Fertility*, **56**, 391–402.
Racey, P.A. (1981). Environmental factors affecting the length of gestation in mammals. In *Environmental Factors in Mammal Reproduction*, pp. 199–213. D. Gilmore and B. Cook (eds). Macmillan, London.
Ralls, K. (1971). Mammalian scent marking. *Science, New York*, **171**, 443–9.
Ralls, K. (1977). Sexual dimorphism in mammals: Avian models and unanswered questions. *American Naturalist*, **111**, 917–38.
Ralls, K., Brugger, K. and Ballou, J. (1979). Inbreeding and juvenile mortality in small populations of ungulates. *Science, New York*, **206**, 1101–3.
Rattray, P.V. (1977). Nutrition and reproductive efficiency. In *Reproduction in Domestic Mammals*, pp. 553–75. H.H. Cole and P.T. Cupps (eds). Academic Press, London.
Redhead, T.D., Enright, N. and Newsome, A.E. (1985). Causes and predictions of outbreaks of *Mus musculus* in irrigated and non-irrigated cereal farms. *Acta Zoologica Fennica*, **173**, 123–7.
Reiter, R.J. (1974). Circannual reproductive rhythms and pineal function: A review. *Chronobiologia*, **1**, 365–95.

Reiter, R.J. (1975). Exogenous and endogenous control of the annual reproductive cycle in the male golden hamster: participation of the pineal gland. *Journal of Experimental Zoology*, **191**, 111–19.
Reiter, R.J. (1980). Seasonal reproduction: An expedient and essential artifice. *Progress in Reproductive Biology*, **5**, 1–4.
Renfree, M.B. and Calaby, J.H. (1981). Background to delayed implantation and embryonic diapause. *Journal of Reproduction and Fertility Supplement*, **29**, 1–9.
Reynolds, J. and Keverne, E.B. (1979). The accessory olfactory system and its role in pheromonally mediated suppression of oestrus in grouped mice. *Journal of Reproduction and Fertility*, **57**, 31–5.
Reynolds, G. and Turkowski, F. (1972). Reproductive variations in the round-tailed ground squirrel as related to winter rainfall. Journal of Mammalogy, **53**, 893–8.
Richmond, M. and Conaway, C.H. (1969). Induced ovulation and oestrus in *Microtus ochrogaster. Journal of Reproduction and Fertility, Supplement*, **6**, 357–76.
Roberts, M.W. and Wolfe, J.L. (1974). Social influences on susceptibility to predation in cotton rats. Journal of Mammalogy, **55**, 869–72.
Rogers, J.G. and Beauchamp, G.K. (1976). Influence of stimuli from populations of *Peromyscus leucopus* on maturation of young. Journal of Mammalogy, **57**, 320–30.
Romesburg, H.C. (1981). Wildlife science: gaining reliable knowledge. *Journal of Wildlife Management*, **45**, 293–313.
Ropartz, P. (1968). Mise en évidence d'une augmentation de l'activité locomotrice des groups de souris femelles en response á l'odeur d'une groupe de males étranges. *Comptes Rendus Hebdominaires des Seances de l'Academie de Sciences*, **267**, 2341–3.
Rosenzweig, M.L. and Abramsky, Z. (1980). Microtine cycles: the role of habitat heterogeneity. *Oikos*, **34**, 141–6.
Rowell, T.E. (1974). The concept of social dominance. *Behavioral Biology*, **11**, 131–54.
Rudge, M.R. (1970). Mother and kid behavior in feral goats (*Capra hircus* L.). *Zietschrift für Tierpsychologie*, **27**, 687–92.
Rusch, D.A. and Reeder, W.G. (1978). Population ecology of Alberta red squirrels. *Ecology*, **59**, 400–20.
Russell, F.C. (1948). Diet in relation to reproduction and viability of the young. Part I. Rats and other laboratory animals. *Commonwealth Bureau of Animal Nutrition. Technical Communication*, No. 16, 99pp.
Sacher, G.A. and Staffeldt, E.F. (1974). Relation of gestation time to brain weight for placental mammals: implications for the theory of vertebrate growth. *American Naturalist*, **108**, 593–615.
Sadleir, R.M.F.S. (1965). The relationships between agonistic behaviour and population changes in the deer mouse, *Peromyscus maniculatus. Journal of Animal Ecology*, **34**, 331–52.
Sadleir, R.M.F.S. (1969). *The Ecology of Reproduction in Wild and Domestic Mammals*. Methuen, London.
Sadleir, R.M.F.S., Casperson, K.D. and Harling, I.J. (1973). Intake and requirements of energy and protein for the breeding of wild deermice, *Peromyscus maniculatus. Journal of Reproduction and Fertility, Supplement*, **19**, 237–52.
St. Amant, J.L.S. (1970). The detection of regulation in animal populations. *Ecology*, **51**, 823–8.
Sanders, E.H., Gardner, P.D., Berger, P.J. and Negus, N.C. (1981). 6-Methoxybenzoxazolinone: A plant derivative that stimulates reproduction in *Microtus montanus. Science, New York*, **214**, 67–9.
Schaffer, M.L. (1978). Determining minimum viable population sizes: A case study of the grizzly bear (*Ursus arctos* L.). Unpublished PhD dissertation, Duke Univer-

sity, Durham, N.C. (Quoted in Schaffer, 1981).
Schaffer, M.L. (1981). Minimum population sizes for species conservation. *BioScience*, **31**, 131–4.
Schaffer, W.M. (1974). Selection for optimal life-histories: the effect of age structure. *Ecology*, **55**, 291–303.
Schally, A.V., Arimura, A. and Kastin, A.J. (1973). Hypothalamic regulatory hormones. *Science, New York*, **179**, 341–50.
Scheffer, V.B. (1951). The rise and fall of a reindeer herd. *Science Monthly*, **73**, 356–62.
Schiller, E.L. (1956). Ecology an health of *Rattus* at Nome, Alaska. *Journal of Mammalogy*, **37**, 181–8.
Schlesinger, W.H. (1976). Toxic foods and vole cycles: additional data. *American Naturalist*, **110**, 315–17.
Schultz, A.M. (1964). The nutrient-recovery hypothesis for arctic microtine cycles II. In *Grazing in Terrestrial and Marine Environments*, pp. 57–68. D.J. Crisp (ed.). Blackwell Scientific Publications, Oxford.
Schnell, J.H. (1968). The limiting effects of natural predation on experimental cotton rat populations. *Journal of Wildlife Management*, **32**, 698–711.
Scott, J.W. and Pfaff, D.W. (1970). Behavioural and electrophysiological responses of female mice to male urine odours. *Physiology and Behaviour*, **5**, 407–11.
Scott, R.S. and Burger, H.G. (1981). Mechanism of action of Inhibin. *Biology of Reproduction*, **24**, 541–50.
Selander, R.K. (1972). Sexual selection and dimorphism in birds. In *Sexual Selection and the Descent of Man*, pp. 180–230. B. Campbell (ed.). Aldine, Chicago.
Selye, H. (1946). The general adaptation syndrome and the diseases of adaptation. *Journal of Clinical Endocrinology*, **6**, 117–230.
Signoret, J.P. (1976). Chemical communication and reproduction in domestic mammals. In *Mammalian Olfaction: Reproductive Processes and Behaviour*, pp. 243–56. R.L. Doty (ed.). Academic Press, New York.
Sinclair, A.R.E. (1973). Regulation, and population models for a tropical ruminant. *East African Wildlife Journal*, **11**, 307–16.
Sinclair, A.R.E. (1974a). The natural regulation of buffalo populations in East Africa. I. Introduction and resource requirements. *East African Wildlife Journal*, **12**, 135–54.
Sinclair, A.R.E. (1974b). The natural regulation of buffalo populations in East Africa. II. Reproduction, recruitment and growth. *East African Wildlife Journal*, **12**, 169–84.
Sinclair, A.R.E. (1974c). The natural regulation of buffalo populations in East Africa. III. Population trends and mortality. *East African Wildlife Journal*, **12**, 185–200.
Sinclair, A.R.E. (1974d). The natural regulation of buffalo populations in East Africa. IV. The food supply as a regulating factor, and competition. *East African Wildlife Journal*, **12**, 291–311.
Sinclair, A.R.E. (1975). The resource limitation of trophic levels in tropical grassland ecosystems. *Journal of Animal Ecology*, **44**, 497–520.
Sinclair, A.R.E. (1981). Environmental carrying capacity and the evidence for overabundance. In *Problems in Management of Locally Abundant Wild Mammals*, pp. 247–57. P.A. Jewell and S. Holt, (eds). Academic Press, New York.
Sinclair, A.R.E., Krebs, C.J. and Smith, J.N.M. (1982). Diet quality and food limitation in herbivores: the case of the snowshoe hare. *Canadian Journal of Zoology*, **60**, 889–97.
Sisson, S. (1975). *The Anatomy of Domestic Animals*, 5th edition, revised by R. Getty. W.B. Saunders & Co., Philadelphia.

Skinner, J.D., Nel, J.A.J. and Millar, R.P. (1977). Evolution of time of parturition and differing litter sizes as an adaptation to changes in environmental conditions. In *Reproduction and Evolution*, pp. 39–44. J.H. Calaby and C.H. Tyndale-Biscoe (eds). Australian Academy of Science, Canberra.

Skutch, A.F. (1949). Do tropical birds rear as many young as they can nourish? *Ibis*, **91**, 430–55.

Skutch, A.F. (1967). Adaptive limitation of the reproductive rate of birds. *Ibis*, **109**, 579–99.

Slater, P.J.B. (1978). *Sex Hormones and Behaviour*. Studies in Biology No. 103. Edward Arnold, London.

Slobodkin, L.B. (1962). *Growth and Regulation of Animal Populations*. Holt, Rinehart and Winston, New York.

Smith, F.E. (1935). The role of biotic factors in the determination of population densities. *Journal of Economic Entomology*, **28**, 873–98.

Smith, J.C. (1981). Senses and communication. *Symposia of The Zoological Society of London*, **47**, 367–93.

Smith, M.H. (1971). Food as a limiting factor in the population ecology of *Peromyscus polionotus* (Wagner). *Annales Zoologici Fennici*, **8**, 109–12.

Smith, M.H., Gentry, J.B. and Pinder, J. (1974). Annual fluctuations in a small mammal population in an eastern hardwood forest. *Journal of Mammalogy*, **55**, 231–4.

Smith M.H., Garten, C.T. and Ramsey, P.R. (1975). Genetic heterozygosity and population dynamics in small mammals. In *Isozymes IV. Genetics and Evolution*, pp. 85–102. C.L. Markert (ed.). Academic Press, New York.

Smith, R.L. (1976). Ecological genesis of endangered species: The philosophy of preservation. *Annual Review of Ecology and Systematics*, **7**, 33–55.

Smuts, G.L. (1978). Interactions between predators, prey and their environment. *BioScience*, **28**, 316–20.

Soriguer, R.C. and Rogers, P.M. (1981). The European wild rabbit in mediterranean Spain. In *Proceedings of the World Lagomorph Conference, Ontario, 1979*, pp. 654–64. K. Myers and C.D. MacInnes (eds). University of Guelph.

Southern, H.N. (1970). The natural regulation of the tawny owl (*Strix aluco*). *Journal of Zoology, London*, **162**, 197–285.

Southern, H.N. and Lowe, V.P.W. (1982). Predation by tawny owls (*Strix aluco*) on bank voles (*Clethrionomys glareolus*) and wood mice (*Apodemus sylvaticus*). *Journal of Zoology, London*, **198**, 83–102.

Southwood, T.R.E. (1975). The dynamics of insect populations. In *Insects, Science and Society*, pp. 151–99. D. Pimentel (ed.). Academic Press, New York.

Spencer, A.W. and Steinhoff, H.W. (1968). An explanation of geographic variation in litter size. Journal of Mammalogy, **49**, 281–6.

Spinage, C.A. (1969). Naturalistic observations on the reproductive and maternal behaviour of the Uganda defassa waterbuck. *Zeitschrift für Tierpsychologie*, **26**, 39–47.

Spinage, C.A. (1973). The role of photoperiodism in the seasonal breeding of tropical ungulates. *Mammal Review*, **3**, 71–84.

Stabenfeldt, G.H. and Hughes, J.P. (1977). *Reproduction in Horses*, 3rd edition, pp. 401–31. H.H. Cole and P.T. Cupps (eds). Academic Press, London.

Stearns, S.C. (1976). Life-history tactics: A review of the ideas. *Quarterly Review of Biology*, **51**, 3–41.

Stehn, R.A. and Richmond, M.E. (1975). Male-induced pregnancy termination in the prairie vole, *Microtus ochorogaster*. *Annals of the New York Academy of Science*, **187**, 1211–13.

Stenseth, N.C. (1977). Evolutionary aspects of demographic cycles: the relevance of

some models of cycles for microtine fluctuations. *Oikos*, **29**, 525–38.
Stenseth, N.C. (1983). Causes and consequences of dispersal in small mammals. In *The Ecology of Animal Movement*, pp. 63–101. I.R. Swingland, and P.J. Greenwood (eds). Clarendon Press, Oxford.
Stetson, M.H. and Watson-Whitmyre, M. (1976). Nucleus suprachiasmaticus: The biological clock in the hamster? *Science, New York*, **191**, 197–9.
Stoddart, D.M. (1980). *The Ecology of Vertebrate Olfaction*. Chapman and Hall Ltd., London.
Sullivan, T. (1977). Demography and dispersal in island and mainland populations of the deer mouse, *Peromyscus maniculatus*. *Ecology*, **58**, 964–78.
Sumption, K.J. and Flowerdew, J.R. (1985). The ecological effects of the decline in rabbits (*Oryctolagus cuniculus*) due to myxomatosis. *Mammal Review*, **15**, 151–86.
Taitt, M.J. (1981). The effect of extra food on small rodent populations: I. Deermice (*Peromyscus maniculatus*). *Journal of Animal Ecology*, **50**, 111–124.
Taitt, M.J. and Krebs, C.J. (1981). The effect of extra food on small rodent populations: II. Voles (*Microtus townsendii*). *Journal of Animal Ecology*, **50**, 125–37.
Taitt, M.J. and Krebs, C.J. (1983). Predation, cover and food manipulation during a spring decline of *Microtus townsendii*. *Journal of Animal Ecology*, **52**, 837–48.
Tamarin, R.H. (1978a). Dispersal, population regulation, and K-selection in field mice. *American Naturalist*, **112**, 545–55.
Tamarin, R.H. (1978b). A defense of single-factor models of population regulation. In *Populations of Small Mammals Under Natural Conditions*, pp. 159–62. D.P. Snyder (ed.). Special Publication Series Vol. 5, Pymatuning Laboratory of Ecology, University of Pittsburgh.
Tamarin, R.H. (1983). Animal population regulation through behavioral interactions. In *Advances in the Study of Mammalian Behavior*, pp. 698–720. J.F. Eisenberg and D.G. Kleiman (eds). Special Publication No. 7 of the American Society of Mammalogists, Pennsylvania.
Tamarin, R.H. and Krebs, C.J. (1969). *Microtus* population biology II. Genetic changes at the transferrin locus in fluctuating populations of two vole species. *Evolution*, **23**, 183–211.
Tamarin, R.H., Reich, L.M. and Moyer, C.A. (1984). Meadow vole cycles within fences. *Canadian Journal of Zoology*, **62**, 1796–804.
Tamarkin, L., Westrom, W., Hamill, A. and Goldman, B. (1976). Effect of melatonin on the reproductive systems of male and female Syrian hamsters: a diurnal rhythm in sensitivity to melatonin. *Endocrinology*, **99**, 1534–41.
Tapper, S.C. (1976). Population fluctuations of field voles (*Microtus*): a background to the problems involved in predicting vole plagues. *Mammal Review*, **6**, 93–117.
Tapper, S.C. (1979). The effect of fluctuating vole numbers (*Microtus agrestis*) on a population of weasels (*Mustela nivalis*) on farm-land. *Journal of Animal Ecology*, **48**, 603–17.
Tapper, S. (1982). Using estate records to monitor population trends in game and predator species, particularly weasels and stoats. *Transactions of the International Congress of Game Biologists*, **14**, 115–21.
Tast, J. and Kalela, O. (1971). Comparison between rodent cycles and plant production in Finnish Lapland. *Annales Academiae Scientarum Fennicae, Series A.IV. Biology*, **186**, 1–14.
Teague, L.G. and Bradley, E.L. (1978). The existence of a puberty accelerating pheromone in the urine of the male prairie deermouse (*Peromyscus maniculatus bairdii*). *Biology of Reproduction*, **19**, 314–17.

Telfer, E.S. (1967). Comparison of moose and deer winter range in Nova Scotia. *Journal of Wildlife Management*, **31**, 418–25.

Thibault, C. (1973). Sperm transport and storage in vertebrates. *Journal of Reproduction and Fertility, Supplement*, **18**, 39–53.

Thibault, C., Courot, M., Martinet, L., Mauleon, P., DuMesnil du Buisson, F., Ortavant, R., Pelletier, J. and Signoret, J.P. (1966). Regulation of breeding season and oestrous cycles by light and external stimuli in some mammals. *Journal of Animal Science*, **25**, (Supplement), 119–142.

Thompson, D.C. (1977). Reproductive behaviour in the grey squirrel. *Canadian Journal of Zoology*, **55**, 1176–84.

Thompson, D.Q. (1955). The role of food and cover in population fluctuations of the brown lemming at Point Barrow. Alaska. *Transactions of the North American Wildlife Conference*, **20**, 166–74.

Thompson, H.V. (1977). Mink *Mustela vison*. In *The Handbook of British Mammals*, pp. 353–7. G.B. Corbet and H.N. Southern (eds). Blackwell Scientific Publications, Oxford.

Thorpe, P.A. and Herbert, J. (1976). The accessory optic system of the ferret. *Journal of Comparative Neurology*, **170**, 295–310.

To, L.P. and Tamarin, R.H. (1977). The relation of population density and adrenal gland weight in cycling and noncycling voles (*Microtus*). *Ecology*, **58**, 928–34.

Toran-Allerand, C.D. (1978). Gonadal hormones and brain development: Cellular aspects of sexual differentiation. *American Zoologist*, **18**, 553–65.

Trivers, R.L. (1972). Parental investment and sexual selection. In *Sexual Selection and the Descent of Man* 1871–1971, pp. 136–79. B. Campbell (ed.). Aldine, Chicago.

Trivers, R.L. (1974). Parent-offspring conflict. *American Zoologist*, **14**, 249–64.

Tuomi, J. (1980). Mammalian reproductive strategies: A generalised relation of litter size to body size. *Oecologia*, **45**, 39–44.

Turek, F.W. and Campbell, C.S. (1979). Photoperiodic regulation of neuroendocrine-gonadal activity. *Biology of Reproduction*, **20**, 32–50.

Turek, F.W. and Losee, S.H. (1978). Melatonin-induced testicular growth in golden hamsters maintained on short days. *Biology of Reproduction*, **18**, 299–305.

Turek, F.W., Alvis, J.D. and Menaker, M. (1977). Pituitary responsiveness to LRF in castrated male hamsters exposed to different photoperiodic conditions. *Neuroendocrinology*, **24**, 140–6.

Turek, F.W., Desjardins, C. and Menaker, M. (1976). Differential effects of melatonin on the testes of photoperiodic and nonphotoperiodic rodents. *Biology of Reproduction*, **15**, 94–7.

Turner, B.N and Iverson, S.L. (1973). The annual cycle of aggression in male *Microtus pennsylvanicus* and its relation to population parameters. *Ecology*, **54**, 967–81.

Turner, C.D. (1966). *General Endocrinology*. W.B. Saunders, Philadelphia.

Twigg, G.I. (1975a). Chapter 1. Finding mammals. *Mammal Review*, **5**, 71–82.

Twigg, G.I. (1975b). Chapter 2. Catching mammals. *Mammal Review*, **5**, 83–100.

Twigg, G.I. (1975c). Chapter 3. Marking mammals. *Mammal Review*, **5**, 101–116.

Tyndale-Biscoe, H. (1973). *Life of Marsupials*. Edward Arnold, London.

Vandenbergh, J.G. (1967). Effect of the presence of a male on the sexual maturation of female mice. *Endocrinology*, **81**, 345–9.

Vandenbergh, J.G. (1971). The influence of the social environment on sexual maturation in male mice. *Journal of Reproduction and Fertility*, **24**, 383–90.

Vandenbergh, J.G. (1973). Acceleration and inhibition of puberty in female mice by pheromones. *Journal of Reproduction and Fertility, Supplement*, **19**, 411–19.

Vandenbergh, J.G. (1976). Acceleration of sexual maturation in female rats by stimulation. *Journal of Reproduction and Fertility*, **46**, 451–3.

Vandenbergh, J.G., Whitsett, J.M. and Lombardi, J.R. (1975). Partial isolation of a pheromone accelerating puberty in female mice. *Journal of Reproduction and Fertility*, **43**, 515–23.

Vandenbergh, J.G., Finlayson, J.S., Dobrogosz, W.J., Dills, S.S. and Kost, T.A. (1976). Chromatographic separation of puberty accelerating pheromone from male mouse urine. *Biology of Reproduction*, **15**, 260–5.

Varley, G. and Gradwell, G.R. (1960). Key factors in population studies. *Journal of Animal Ecology*, **29**, 399–401.

Varley, G. and Gradwell, G.R. (1968). Population models for the winter moth. In *Insect Abundance, Symposia of the Royal Entomological Society of London, No. 4*, pp. 132–42. T.R.E. Southwood (ed.). Blackwell Scientific Publications, Oxford.

Vaughan, T.A. (1969). Reproduction and population densities in a montane small mammal fauna. *University of Kansas Museum of Natural History Miscellaneous Publications*, **51**, 51–74.

Verner, J. and Willson, M.F. (1966). The influence of habitats on mating systems of North American passerine birds. *Ecology*, **47**, 143–47.

Wagner, F.H. and Stoddart, L.C. (1972). Influence of coyote predation on black-tailed jackrabbit populations in Utah. *Journal of Wildlife Management*, **36**, 329–42.

Walton, A. (1960). Copulation and natural insemination. In *Marshall's Physiology of Reproduction Vol. I. Part II*, 3rd edition, pp. 130–60. A.S. Parkes (ed.). Longmans, Green & Co., London.

Watson, A. and Moss, R. (1970). Dominance, spacing behaviour and aggression in relation to population limitation in vertebrates. In *Animal Populations in Relation to Their Food Resources*, pp. 167–218. A. Watson (ed.). Blackwell Scientific Publications, Oxford.

Watts, C.H.S. (1969). The regulation of wood mouse (*Apodemus sylvaticus*) numbers in Wytham woods, Berkshire. *Journal of Animal Ecology*, **38**, 285–304.

Wells, L.J. and Zalesky, M. (1940). Effects of low temperature environment on the reproductive organs of male mammals with annual aspermia. *American Journal of Anatomy*, **66**, 429–38.

Weir, B.J. (1971). The reproductive organs of the female plains viscacha, *Lagostomus maximus*. *Journal of Reproduction and Fertility*, **25**, 365–73.

Weir, B.J. (1973). The induction of ovulation and oestrus in the chinchilla. *Journal of Reproduction and Fertility*, **33**, 61–8.

Weir, B.J. and Rowlands, I.W. (1973). Reproductive strategies of mammals. *Annual Review of Ecology and Systematics*, **4**, 139–63.

Wetterberg, L., Geller, E. and Yuwilder, A. (1970). Harderian gland: an extraretineal photo receptor influencing the pineal gland in neonatal rats? *Science, New York*, **167**, 884–5.

White, I.G., Roger, J.C., Morris, S.R. and Marley, P.B. (1977). The role of secretions of the male accessory organs of mammals. In *Reproduction and Evolution*, pp. 183–92. J.H. Calaby, C.H. Tyndale-Biscoe (eds). Australian Academy of Science, Canberra.

Wiegert, R.G. (1972). Avian versus mammalian predation on a population of cotton rats. *Journal of Wildlife Management*, **36**, 1322–7.

Whitten, W.K. (1956). Modifications of the oestrous cycle of the mouse by external stimuli associated with the male. *Journal of Endocrinology*, **13**, 399–404.

Whitten, W.K. (1958). Modifications of the oestrous cycle of the mouse by external stimuli associated with the male. Changes in the oestrous cycle determined by vaginal smears. *Journal of Endocrinology*, **17**, 307–13.

Whitten, W.K. (1959). Occurrence of anoestrus in mice caged in a group. *Journal of Endocrinology*, **18**, 102–7.

Whitten, W.K. (1966). Pheromones and mammalian reproduction. *Advances in Reproductive Physiology*, **1**, 155–77.

Widdowson, E.M. and Cowan, J. (1972). The effect of protein deficiency and calorie deficiency on the reproduction of rats. *British Journal of Nutrition*, **25**, 85–95.

Wight, H.M. and Conaway, C.H. (1961). Weather influences on the onset of breeding in Missouri Cottontails. *Journal of Wildlife Management*, **25**, 87–9.

Wilbur, H.M., Tinkle, D.W. and Collins, J.P. (1974). Environmental certainty, trophic level, and resource availability in life history evolution. *American Naturalist*, **108**, 805–17.

Williams, G.C. (1966). *Adapatation and Natural Selection*. Princeton University Press, Princeton.

Wilson, C.A. (1979). Hypothalamic neurotransmitters and gonadotrophin release. In *Oxford Reviews of Reproductive Physiology*, Vol. 1, pp. 383–473. C.A. Finn (ed.). Clarendon Press, Oxford.

Wilson, E.O. (1975). *Sociobiology: The New Synthesis*. Harvard University Press, Cambridge, Mass.

Wilson, E.O. and Bossert, W.H. (1963). Chemical communication among animals. *Recent Progress in Hormone Research*, **19**, 673–716.

Wilson, J.D. (1978). Sexual differentiation. *Annual Review of Physiology*, **40**, 279–306.

Wilson, J.D., Griffin, J.E. and George, F.W. (1980). Sexual differentiation: Early hormone synthesis and action. *Biology of Reproduction*, **22**, 9–17.

Wilsson, L. (1971). Observations and experiments on the ecology of the European beaver (*Castor fiber* L.). *Viltrevy*, **8**, 115–266.

Wimsatt, W.A. (1960). Some problems of reproduction in relation to hibernation in bats. *Bulletin of the Museum of Comparative Zoology, Harvard*, **124**, 249–70.

Wimsatt, W.A. (1975). Some comparative aspects of implantation. *Biology of Reproduction*, **12**, 1–40.

Winans, S.S. and Powers, J.B. (1977). Olfactory and vomeronasal deafferentiation of male hamsters: Histological and behavioral analyses. *Brain Research*, **126**, 325–44.

Windberg, L.A. and Keith, L.B. (1976). Experimental analysis of dispersal in snowshoe hare populations. *Canadian Journal of Zoology*, **54**, 2061–81.

Winters, G.H. (1978). Production, mortality and sustainable yield of Northwest Atlantic harp seals (*Pagophilus groenlandicus*). *Journal of the Fisheries Research Board of Canada*, **35**, 1249–61.

Wise, P,M., Rance, N., Barr, G.D. and Barraclough, C.A. (1979). Further evidence that luteinizing hormone releasing hormone also is follicle stimulating hormone releasing hormone. *Endocrinology*, **104**, 940–7.

Wittenberger, J.F. (1976). The ecological factors selecting for polygyny in birds. *American Naturalist*, **110**, 779–99.

Wood, D.H. (1980). The demography of a rabbit population in an arid region of New South Wales, Australia. *Journal of Animal Ecology*, **49**, 55–79.

Wolff, J.O. (1980). The role of habitat patchiness in the population dynamics of snowshoe hares. *Ecological Monographs*, **50**, 111–30.

Wolton, R.J. (1985). The ranging and nesting behaviour of wood mice, *Apodemus sylvaticus* (Rodentia: Muridae), as revealed by radio-tracking. *Journal of Zoology, London, (A)*, **206**, 203–24.

Wynne-Edwards, V.C. (1962). *Animal Dispersion in Relation to Social Behaviour*. Oliver and Boyd, Edinburgh.

Wysocki, C.J. (1979). Neurobehavioural evidence for the involvement of the vomeronasal system in mammalian reproduction. *Neuroscience and Biobehavioral Reviews, 3*, 301–41.

Wysocki, C.J., Wellington, J.L. and Beauchamp, G.K. (1980). Access of urinary non-volatiles to the mammalian vomeronasal organ. *Science, New York*, **207**, 781–3.
Yen, S.C.C. (1977). Regulation of the hypothalamic-pituitary-ovarian axis in women. *Journal of Reproduction and Fertility*, **51**, 181–91.
Zucker, I, Johnston, P.G. and Frost, D. (1980). Comparative physiological and biochronometric analyses of rodent seasonal reproductive cycles. *Progress in Reproductive Biology*, **5**, 102–33.

Indices

General Index

Taxonomic Index